Lecture Notes in Computer Science　　11453

Commenced Publication in 1973
Founding and Former Series Editors:
Gerhard Goos, Juris Hartmanis, and Jan van Leeuwen

Editorial Board Members

David Hutchison
　Lancaster University, Lancaster, UK
Takeo Kanade
　Carnegie Mellon University, Pittsburgh, PA, USA
Josef Kittler
　University of Surrey, Guildford, UK
Jon M. Kleinberg
　Cornell University, Ithaca, NY, USA
Friedemann Mattern
　ETH Zurich, Zurich, Switzerland
John C. Mitchell
　Stanford University, Stanford, CA, USA
Moni Naor
　Weizmann Institute of Science, Rehovot, Israel
C. Pandu Rangan
　Indian Institute of Technology Madras, Chennai, India
Bernhard Steffen
　TU Dortmund University, Dortmund, Germany
Demetri Terzopoulos
　University of California, Los Angeles, CA, USA
Doug Tygar
　University of California, Berkeley, CA, USA

More information about this series at http://www.springer.com/series/7407

Anikó Ekárt · Antonios Liapis ·
María Luz Castro Pena (Eds.)

Computational Intelligence in Music, Sound, Art and Design

8th International Conference, EvoMUSART 2019
Held as Part of EvoStar 2019
Leipzig, Germany, April 24–26, 2019
Proceedings

 Springer

Editors
Anikó Ekárt
Aston University
Birmingham, UK

Antonios Liapis
University of Malta
Msida, Malta

María Luz Castro Pena
University of A Coruña
A Coruña, Spain

ISSN 0302-9743   ISSN 1611-3349   (electronic)
Lecture Notes in Computer Science
ISBN 978-3-030-16666-3   ISBN 978-3-030-16667-0   (eBook)
https://doi.org/10.1007/978-3-030-16667-0

Library of Congress Control Number: 2019936145

LNCS Sublibrary: SL1 – Theoretical Computer Science and General Issues

This Springer imprint is published by the registered company Springer Nature Switzerland AG
The registered company address is: Gewerbestrasse 11, 6330 Cham, Switzerland

Preface

EvoMUSART 2019—the 8th International Conference on Computational Intelligence in Music, Sound, Art and Design—took place during April 24–26, 2019, in Leipzig, Germany, as part of the evo* event.

Following the success of previous events and the importance of the field of computational intelligence—specifically, evolutionary and biologically inspired (artificial neural network, swarm, alife) music, sound, art and design—evoMUSART has become an evo* conference with independent proceedings since 2012.

The use of computational intelligence for the development of artistic systems is a recent, exciting, and significant area of research. There is a growing interest in the application of these techniques in fields such as: visual art and music generation, analysis, and interpretation; sound synthesis; architecture; video; poetry; design; and other creative tasks.

The main goal of evoMUSART 2019 was to bring together researchers who are using computational intelligence techniques for artistic tasks, providing the opportunity to promote, present, and discuss ongoing work in the area. As always, the atmosphere was fun, friendly, and constructive.

EvoMUSART has grown steadily since its first edition in 2003 in Essex, UK, when it was one of the Applications of Evolutionary Computing workshops. Since 2012 it has been a full conference as part of the Evo* co-located events.

EvoMUSART 2019 received 24 submissions. The peer-review process was rigorous and double-blind. The international Program Committee, listed herein, was composed of 47 members from 17 countries. EvoMUSART continued to provide useful feedback to authors: among the papers sent for full review, there were on average 3.3 reviews per paper. The number of accepted papers was 11 long talks (46% acceptance rate) and five posters accompanied by short talks, yielding an overall acceptance rate of 66%.

As always, the EvoMUSART proceedings cover a wide range of topics and application areas, including: generative approaches to music and visual art; deep learning; evolutionary games; and culinary arts. This volume of proceedings collects the accepted papers.

As in previous years, the standard of submissions was high, and good-quality papers had to be rejected. We thank all authors for submitting their work, including those whose work was not accepted for presentation on this occasion.

The work of reviewing is done voluntarily and generally with little official recognition from the institutions where reviewers are employed. Nevertheless, professional reviewing is essential to a healthy conference. Therefore, we particularly thank the members of the Program Committee for their hard work and professionalism in providing constructive and fair reviews.

EvoMUSART 2019 was part of the Evo* 2019 event, which included three additional conferences: EuroGP 2019, EvoCOP 2019, and EvoApplications 2019. Many people helped to make this event a success.

We thank the invited keynote speakers, Manja Marz (Friedrich Schiller University Jena, Germany) and Risto Miikkulainen (University of Texas, USA), for their inspirational talks.

We thank SPECIES, the Society for the Promotion of Evolutionary Computation in Europe and Its Surroundings, for its sponsorship and HTWK Leipzig University of Applied Sciences for their patronage of the event.

We thank the local organizing team lead by Hendrik Richter, and the Faculty of Electrical Engineering and Information Technology of HTWK Leipzig University of Applied Sciences for supporting the local organization.

We thank Marc Schoenauer (Inria Saclay, France), for continued assistance in providing the MyReview conference management system, and Pablo García Sánchez (University of Cádiz, Spain) for taking care of the Evo* publicity, website, and social media service.

Finally, and above all, we would like to express our most heartfelt thanks to Anna Esparcia-Alcázar (Universitat Politècnica de València, Spain), for her dedicated work and coordination of the event. Without her work, and the work of Jennifer Willies in the past years, Evo* would not enjoy its current level of success as the leading European event on bio-inspired computation.

April 2019

Anikó Ekárt
Antonios Liapis
Luz Castro

Organization

EvoMUSART 2019 was part of Evo* 2019, Europe's premier co-located events in the field of evolutionary computing, which also included the conferences EuroGP 2019, EvoCOP 2019, and EvoApplications 2019.

Organizing Committee

Conference Chairs

Anikó Ekárt	Aston University, UK
Antonios Liapis	University of Malta, Malta

Publication Chair

Luz Castro	University of Coruña, Spain

Program Committee

Peter Bentley	University College London, UK
Tim Blackwell	Goldsmiths College, University of London, UK
Andrew Brown	Griffith University, Australia
Adrian Carballal	University of A Coruña, Spain
Amilcar Cardoso	University of Coimbra, Portugal
Peter Cariani	University of Binghamton, USA
Vic Ciesielski	RMIT, Australia
Kate Compton	University of California Santa Cruz, USA
João Correia	University of Coimbra, Portugal
Palle Dahlstedt	Göteborg University, Sweden
Hans Dehlinger	Independent Artist, Germany
Eelco den Heijer	Vrije Universiteit Amsterdam, The Netherlands
Alan Dorin	Monash University, Australia
Arne Eigenfeldt	Simon Fraser University, Canada
José Fornari	NICS/Unicamp, Brazil
Marcelo Freitas Caetano	IRCAM, France
Philip Galanter	Texas A&M College of Architecture, USA
Andrew Gildfind	Google, Inc., Australia
Scot Gresham Lancaster	University of Texas, Dallas, USA
Carlos Grilo	Instituto Politécnico de Leiria, Portugal
Andrew Horner	University of Science and Technology, Hong Kong, SAR China
Colin Johnson	University of Kent, UK
Daniel Jones	Goldsmiths, University of London, UK
Amy K. Hoover	University of Central Florida, USA

Maximos Kaliakatsos-Papakostas	Aristotle University of Thessaloniki, Greece
Matthew Lewis	Ohio State University, USA
Alain Lioret	Paris 8 University, France
Louis Philippe Lopes	University of Malta, Malta
Roisin Loughran	University College Dublin, Ireland
Penousal Machado	University of Coimbra, Portugal
Tiago Martins	University of Coimbra, Portugal
Jon McCormack	Monash University, Australia
Eduardo Miranda	University of Plymouth, UK
Nicolas Monmarché	University of Tours, France
Marcos Nadal	University of the Balearic Islands, Spain
Michael O'Neill	University College Dublin, Ireland
Somnuk Phon-Amnuaisuk	Brunei Institute of Technology, Malaysia
Douglas Repetto	Columbia University, USA
Juan Romero	University of A Coruña, Spain
Brian Ross	Brock University, Canada
Jonathan E. Rowe	University of Birmingham, UK
Antonino Santos	University of A Coruña, Spain
Marco Scirea	IT University of Copenhagen, Denmark
Benjamin Smith	Indianapolis University, Purdue University, USA
Stephen Todd	IBM, UK
Paulo Urbano	Universidade de Lisboa, Portugal
Anna Ursyn	University of Northern Colorado, USA

Contents

Deep Learning Concepts
for Evolutionary Art

Fazle Tanjil and Brian J. Ross[✉]

Department of Computer Science, Brock University,
1812 Sir Isaac Brock Way, St. Catharines, ON L2S 3A1, Canada
fazle.tanjil@gmail.com, bross@brocku.ca
http://www.cosc.brocku.ca/~bross/

Abstract. A deep convolutional neural network (CNN) trained on millions of images forms a very high-level abstract overview of an image. Our primary goal is to use this high-level content information a given target image to guide the automatic evolution of images using genetic programming. We investigate the use of a pre-trained deep CNN model as a fitness guide for evolution. Two different approaches are considered. Firstly, we developed a heuristic technique called Mean Minimum Matrix Strategy (MMMS) for determining the most suitable high-level CNN nodes to be used for fitness evaluation. This pre-evolution strategy determines the common high-level CNN nodes that show high activation values for a family of images that share an image feature of interest. Using MMMS, experiments show that GP can evolve procedural texture images that likewise have the same high-level feature. Secondly, we use the highest-level fully connected classifier layers of the deep CNN. Here, the user supplies a high-level classification label such as "peacock" or "banana", and GP tries to evolve an image that maximizes the classification score for that target label. Experiments evolved images that often achieved high confidence scores for the supplied labels. However, the images themselves usually display some key aspect of the target required for CNN classification, rather than the entire subject matter expected by humans. We conclude that deep learning concepts show much potential as a tool for evolutionary art, and future results will improve as deep CNN models are better understood.

Keywords: Deep convolutional neural network ·
Genetic programming · Evolutionary art

1 Introduction

Evolutionary art has been a subject of attention for many years [1–4]. Due to the exploratory nature of evolutionary computation, evolutionary methodologies have been of great interest in art and design applications [5,6]. Early evo-art systems were interactive, in which fitness is manually assigned a score based on visual appeal [7]. Later, automatic techniques for fitness evaluation

© Springer Nature Switzerland AG 2019
A. Ekárt et al. (Eds.): EvoMUSART 2019, LNCS 11453, pp. 1–17, 2019.
https://doi.org/10.1007/978-3-030-16667-0_1

were explored, and a recent goal is to model the aesthetics of images [8]. One of the major challenges in automating evolutionary art is how to intelligently guide the high-level characteristics of evolved images. This is an important aspect of creative image generation because, even for human artists, high-level representations are critical. Finding new ways to give higher-level control to the evolutionary art system continues to be an active research topic.

In the past, the idea of having a high-level abstract overview of a target object image has been difficult to realize. Although many computer vision techniques have been used to evaluate low-level features of an image, these approaches are usually *ad hoc* solutions that do not consider high-level image content. On the other hand, recent breakthroughs in deep artificial neural networks [9–12] are of great potential in addressing this problem [12]. A deep convolutional neural network (CNN) trained on millions of images forms a very high-level abstract overview of image content [12]. A deep CNN is capable of having a high-level abstract overview of an image regardless of the low-level pixel positions or spatial information. It can also retain spatial invariance, which means changes to size and position do not affect the high-level representation.

This paper investigates the integration of an evolutionary generative art system with a deep CNN. The deep CNN will serve as a fitness guide to the evolutionary system. In doing this, we do not perform any training, but instead, we use a pre-trained deep CNN designed for object recognition. The approach we use is to first pass a target object image (or content image) to the deep CNN. The CNN will form a high-level overview of that image, and the CNN behaviour on that image will be used as a fitness target for the genetic programming-based evo-art system. This way, the evolutionary system be guided by the content image during its exploration of images.

In the past, there were few ways to represent the high-level abstraction of images. Popular computer vision algorithms like SIFT [13], HOG [14], and SURF [15] detect the local features like a key point, an edge, or a blob, but fail to capture overall high-level image representations. They essentially work as feature detectors and ordinarily perform poor generalization. On the other hand, a deep CNN uses several hidden layers to hierarchically model the high-level representation of an image. For example, the first layer might detect edges in the image. The second layer might detect corners present in the content image based on the previous layer's detected edges. The third layer detects other features, and so on. Changes in one feature do not drastically change the whole representation of an image in the higher layers. Each deep layer captures characteristics of multiple features of the previous layer. The ability to model vast combinations of data in a hierarchical and compressed manner is a strength of deep neural networks.

Our research uses a pre-trained CNN system within a genetic programming-based evo-art system. We propose a heuristic for determining which components of the deep CNN are most effective to use for image fitness evaluation. Additionally, we investigate the use of the final classification layer of the CNN system to guide image evolution. The classification layer is the most abstract representation of the target object within the deep CNN. We perform these investigations using the popular VGG (Visual Geometry Group) deep CNN system, which is pre-trained to recognize 1000 different categories of objects [9]. Hence, we did not perform any CNN training during our research.

The paper is organized as follows. Section 2 provides relevant background information and related research. Section 3 describes the architecture of our hybrid GP/deep CNN system, our heuristic strategy for determining which components of the CNN to use in image evaluation, and experimental results. Section 4 explores the use of the fully connected layer of the CNN for fitness evaluation. Section 5 gives concluding remarks and future research directions.

Further details of this research are in [16]. The paper presumes familiarity with genetic programming [17,18] and procedural image evolution [7].

2   Background

2.1   Deep Learning and Convolutional Neural Networks

Generally speaking, deep learning allows computers to learn from experience and understand the world in terms of a hierarchy of concepts, with each concept defined in terms of its relation to simpler concepts [19]. If we draw a graph showing how these concepts are built on top of each other, the graph is *deep*, with many layers (see Fig. 1). The basis of deep learning architectures is derived from artificial neural networks, and the concept of a deep hierarchy is denoted by multiple hidden layers of the network. A feedforward deep neural network is the most common example of deep learning.

CNNs are a specialized form of deep neural networks [19]. Like other neural networks, CNNs have neurons that have learnable weights and biases. Each neuron receives inputs, performs some operation and based on the architecture, and sends it to other neurons. However, CNNs differ from regular neural networks in a few ways. A CNN's input data usually has a grid-like shape, for example, a bitmap image. Due to this grid topology, the CNN encodes specific characteristics into the architecture. For example, CNNs use *pooling layers*, which downsample the data, thereby reducing the dimensionality and improving efficiency. The use of *convolution operators* is also unique to the CNN. Convolutions in the network process grid-shaped inputs from input nodes, resulting in *feature maps* (or *activation maps*) of the processed input. Convolution operations take the form of matrix operators. In the case of bitmap data, convolutions are image filters. Low-level convolution filters may denote intuitive operations such as edge detection and noise reduction. Higher level convolutions become more abstract. The operators themselves are usually altered during training.

2.2   CNNs, Evolution, and Art

Our CNN architecture is inspired by work by Gatys *et al.* for image style transfer [12]. Two images are given to their system – a style image, and a content image. The content image is then re-rendered using the image characteristics found in the style image. The results are remarkable, and their approach represents an impressive solution to a much studied classical research problem in computer graphics – stylized non-photorealistic rendering [20]. Rather than hand-derive a rendering algorithm that produces a single rendering effect, their system is capable of faithfully learning and reproducing virtually any artistic style.

Another relevant neural network paper is by Nguyen *et al.* [21]. They use an evolutionary algorithm to generate images that are then passed through a deep neural network. Their primary goal is to fool the DNN's classification abilities with the evolved images. The evolved images indeed have high prediction score from the DNN, and yet may be unrecognizable to humans.

Other examples using evolution and neural networks in art applications are as follows. Bontrager *et al.* combine generative adversarial networks (GANs) with interactive evolutionary computation [22]. They show that a GAN trained on a specific target domain can act as an efficient genotype-to-phenotype mapping. A trained GAN, the vector given as input to the GAN's generator network can be controlled using evolutionary techniques, allowing high-quality image generation. Stanley uses an evolution-based neural network called NEAT for image generation [23]. By interactively evolving networks, a spaceship design (for example) can be evolved. Beluja *et al.* use artificial neural nets to learn the user's aesthetic preferences [24]. A paper showing the benefit of integrating GP and deep learning is by Agapitos *et al.* [25], in which GP is used to evolve a multi-layer image classifier for handwritten digits.

3   Evolution Using CNN Feature Maps

3.1   System Design

Our system is inspired by the deep CNN architecture used by Gatys *et al.* [12]. We follow their approach to selecting a high level of the CNN system to model the content image. However, beyond these similarities, the systems fundamentally differ with respect to image synthesis. Gatys *et al.* neural architecture permits pixel-level alteration of the synthesized image, which occurs during back propagation and error minimization. This permits highly accurate re-renderings of the content image, guided by minimizing the style image's error gradient. With our hybrid GP–CNN system, rendering entirely depends upon the evolved procedural image formula. The evolved formula renders an entire canvas, which is then evaluated to obtain a fitness score. It is not possible to easily fine-tune the formula to edit erroneous pixels on the canvas in order to improve fitness. Therefore, it is infeasible for GP-based systems such as ours to perform style transfers like the Gatys *et al.* system.

Figure 1 shows the system architecture. Like Gatys *et al.*, the deep CNN model we use is the VGG system [9]. This is a pre-trained deep CNN system for object classification. We will use the feature space provided by a normalized version of the 16 convolutional and 5 pooling layers of the 19-weight layers VGG-19 network [9]. The model is pre-trained using the ILSVRC 2012 dataset.

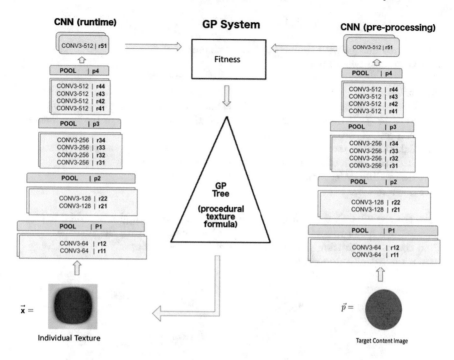

Fig. 1. System Architecture. The CNN on left side is integrated with GP fitness. The same CNN is used (right side) to pre-evaluate a target image, to be used as a basis for determining feature map values to match with evolved images during fitness.

There are several configurations of the VGG architecture introduced in [9]. We use configuration **E**, which is the deepest architecture compared to the other configurations. It has 16 weight convolution layers and 3 weight fully connected layers totaling 19 weight layers.

We use a 256×256 RGB image as an input to the network. Initially, the image goes through a normalization preprocessing stage. This normalized image is then processed through a series of 16 convolution layers. Although the original VGG architecture used max pooling in their pooling layer, we follow [12] and change it to average pooling for better results. We will also change the pooling window size and stride (discussed below).

The deep CNN's target content representation at level r51 is used for fitness to our GP system. However, the high dimensionality (32×32) of activation matrices at this level is too large to use for fitness. Therefore, we altered the pooling layers to downsample the data by changing the window size and stride of the pooling operation from 2 to 3. This results in a matrix of size 3×3, which is more amenable for use by the GP fitness.

Table 1. MMMS algorithm

– Select three images with similar content or shapes.
– Pair them into two groups.

Repeat
    1. Pass each image separately from the same group through the deep CNN.
    2. At layer #r51 take the minimum of two activation matrices of the two images
       respectively from the same group.
    3. Calculate the mean of the resultant minimum matrix from the previous step.
    4. Store the mean of all the resultant minimum activation matrix in
       descending order along with their activation #.
Until no groups are left

– Take the top 40 activations from the sorted values from both groups stored in step 4.
– Divide them into two levels each with 20 activations along with their means.
– Find common activation # in the same level from both groups.
Return the list of common activation #'s.

(a) Pair 1.   (b) Pair 2.

Fig. 2. Target images grouped into pairs.

3.2 Feature Map Selection and the MMMS Heuristic

Our next task is to reduce the number of feature maps. There are 512 feature maps at level r51, which is too many to use, and ultimately unnecessary.

Highly activated neurons are one means towards understanding CNNs. Activated neurons are responding to specific visual attributes present in an image. Our goal is to determine the feature maps that are active when images are given to them. Given a set of target images sharing some desirable image characteristics, by examining the activation levels of feature maps for these images, we can determine the active feature maps common for them. These activation maps are selected for use during fitness evaluation, and the other maps are ignored.

To determine the most promising feature maps to use, we use a heuristic that we call the Mean Min Matrix Strategy (MMMS) (Table 1). This preprocessing is performed before performing a run with the GP-CNN system, in order to predetermine the activation maps to use in subsequent GP runs. Thus, MMMS should be performed for each new target image used for a GP run.

Looking at Table 1, we first need a set of exemplar target (content) images that share common features of interest. We used 3 such images per experiment; one of these images will also be used as a target image during the GP run.

We arrange the set of images of similar objects paired into two groups. The three images are then divided into pairs of images (Fig. 2). Next, we pass each image from a pair (e.g. Pair 1) separately through the deep CNN and extract the 512 activation matrices from CNN level r51. For each image activation, we have two matrices, one for each image of the pair. Consider, the activation matrix #38 from r51 from both of the images:

$$Red\ circle: \begin{bmatrix} 43.7 & 9.8 & 19.1 \\ 16.2 & 47.4 & 0.0 \\ 74.2 & 63.5 & 38.9 \end{bmatrix} \quad Black\ circle: \begin{bmatrix} 61.8 & 42.0 & 39.7 \\ 28.5 & 63.1 & 5.2 \\ 58.2 & 64.4 & 29.2 \end{bmatrix}$$

Next, we apply the MMMS operation. It compares each corresponding matrix element from the two matrices and produces another matrix which contains the minimum value of the two elements:

$$R_{ij} = min(CirRed_{ij}, CirBlack_{ij})\ (1 \leq i, j \leq 3)$$

We minimize the values so that later results are not unduly influenced by spikes caused by artifacts within one image. For the above matrices, the resultant matrix looks like this:

$$Min: \begin{bmatrix} 43.7 & 9.8 & 19.1 \\ 16.2 & 47.4 & 0.0 \\ 58.2 & 63.5 & 29.2 \end{bmatrix}$$

Next, we find the mean of the minimum matrix. This will denote the overall activation level of the activation map. We save the means of all 512 activation maps and sort them in descending order. We do the same with Pair 2 (red circle and distorted circle). Now we have two lists of mean activation values sorted in descending order. Our goal is to select top activations that are common in both groups. The top 40 activations are divided into two levels or batches. The intuition behind having two levels is to make sure that when we try to match between these 2 groups and find the common activations, the position of the common activation is as close as possible. In other words, an activation that is in level 1 in group 1 should also be in level 1 in group 2. By finding the common strong activations, we reduce the 512 feature maps down to a handful that share common activations for images of interest. Although the number will vary according to the images used, we typically identified around 20 activation matrices in our experiments.

3.3   Fitness and Genetic Programming Setup

Once the MMMS has identified a set A of activation maps, fitness will use these maps when evaluating evolved images. Before a GP run, a target image is given to the deep CNN (right hand side of Fig. 1). The activation maps corresponding to the set A are saved for the target image. During the run, fitness evaluation matches a candidate image's activation values on the maps in A with the saved target values using mean square error:

$$Fitness = \sum_{a \in A} (\sum_{i=1}^{3} \sum_{j=1}^{3} (P_{i,j}^a - T_{i,j}^a)^2 / 9)$$

where a indexes the pre-selected activation maps in set A, P^a is a candidate image's activation map, and T^a is the target map.

Table 2. GP parameters

Parameter	Value
Runs per experiment	10
Generations	35
Population size	300
Elitism	1
Tournament size	3
Crossover rate	60%
Maximum crossover depth	17
Mutation	30%
ERC mutation	10%
Maximum mutation depth	17
Prob. of terminals in cross/mutation	10%
Initialization method	Ramped half & half
Tree grow max/min	[5, 5]
Grow probability	50%
Tree half max/min	[2,6]

The core GP system is from [26], from which we use the same ECJ v.23 GP engine [27], texture language, and GP parameters. Table 2 shows the GP parameters used for our experiments. The texture generating language is fairly conventional, and includes terminals (Cartesian coordinates, polar coordinates, ephemerals), mathematical operators, a conditional, and noise generators (turbulence, Perlin, and others).

3.4   Other Software and Hardware Details

The deep learning framework we will use for the implementation of the VGG-19 is PyTorch v.0.2.0 [28], which is based on Python [29]. PyTorch is an optimized tensor library for deep learning using GPUs and CPUs, and runs on CUDA-based PCs. We performed all runs on a NVIDIA TITAN Xp (Compute Capability 6.1) with 12 GB memory. Our CUDA version is 9.0.176. Using CUDA results in about 5-times acceleration in performance compared to CPU-based runs. The GP system (ECJ) runs on 2.80 GHz Intel(R) Core(TM) i7 CPU with 6GB memory, running Ubuntu 16.04.3.

3.5   Single Target Results

Figure 3 shows different target images their 3 best scoring results (selected from 10 runs), along with their mean squared error (MSE) scores. Visually, it is evident that images are capturing characteristics to varying degrees of corresponding target images. The first 4 targets comprising simple geometries derived good results, in which the target image characteristics are evident in all the solutions. The last 2 targets (rings, spiral) were more challenging, and this is reflected in the higher MSE scores. However, even these difficult cases show results that exhibit fundamental characteristics of the target images – nested arcs and curves.

Figure 4 shows fitness performance graphs for the circle and rings experiments. Note the different scales for the fitness axes in the graphs. The fitness plots show that further convergence is possible, and that the maximum generations could be increased. However, we kept the generations low due to computation time.

There are multiple factors that influence the results shown. Firstly the MMMS heuristic used to select promising activation maps is fallible. More accurate results might be obtained if additional and different activation maps were used. Since MMMS relies on the most active maps, other maps with low to medium activation levels are ignored, and such maps can still convey important information about an image. Another factor is the natural bias of the texture formulae evolved by GP. For example, the circle target often generated circular results, which benefit with the GP language's polar coordinates. The other geometric shapes are also easily managed. Circles and spirals, although codable in the texture language, are more difficult expressions to evolve. One other important factor is the nature of the pre-trained VGG network. Although we did not examine the million-plus images used to train it, it is likely the case that the first four simple geometries are fundamental image characteristics of many training images. On the other hand, concentric circles and spirals are more esoteric, and we imagine that these images are foreign to the VGG system, being irrelevant to the training done on the system.

With respect to art applications, it is promising that the different evolved solutions evoke important characteristics of the target images, yet are notably *different* from the targets. This shows that the system is not inhibited by a requirement to duplicate the target images, but has the latitude to produce other visual artifacts, including variants of shape, colour, and other transformations. The CNN guides evolution, but still permits creative and interesting results.

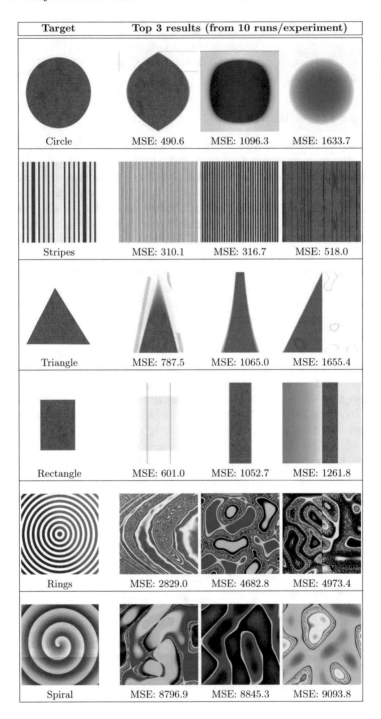

Fig. 3. Results from MMME experiments. One experiment per row. Mean Squared Error given per solution (lower values preferred).

3.6   Multiple Target Results

We performed a few experiments using a combination of two target images. We were curious whether evolved results would capture aspects of both targets. For example, separate images consisting of vertical stripes and horizontal stripes were used as target images. To combine them, we first used MMMS to identify the active feature maps for each of these images. Although there were some duplicate maps, usually the maps for each image were distinct. We then combined the top 50% of each image's maps into one set of feature maps. This combination of feature maps is then used by fitness. All other aspects of the system are the same as before.

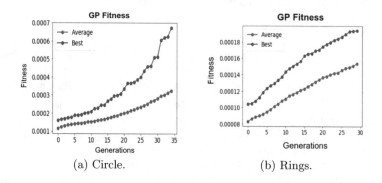

(a) Circle.   (b) Rings.

Fig. 4. Fitness performance. Averaged over 10 runs.

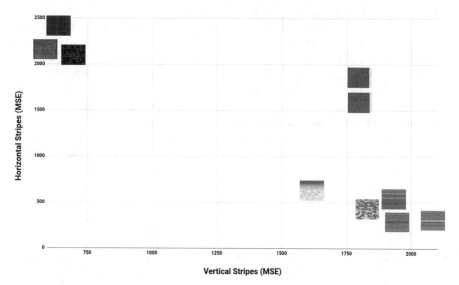

Fig. 5. Scatter plot for horizontal and vertical stripe multiple targets. Each axis measures MSE for the MMMS activation maps for that target image.

The results of combined target images were less successful than the single target runs. Evolved images typically conform to aspects of one of the two target images. Figure 5 shows a scatter plot for the horizontal/vertical stripe experiment, and solutions clearly score higher in either the horizontal or vertical measurements. Since there are usually few feature maps common to both targets, the use of mutually distinct sets of maps explains the results we obtained – solutions will tend to fit one of the two sets of maps. Moreover, the CNN itself is not trained to coalesce features from distinct images, but rather, is used to extract particular visual features required to classify specific subject matter. In conclusion, more work is required to successfully use multiple targets in our system.

4   Evolution Using Fully Connected Layer

4.1   System Design

We will now use the highest layer of a CNN as a fitness guide for image evolution. The goal is to obtain images with similar visual features to subject matter belonging to a supplied class label. For example, if we specify the class label "peacock", which is one of the classes recognized in the output layer of the CNN, our naive intention is to evolve an image with similar characteristics to a peacock. Due to this high level of abstraction, we no longer need to consider the direct use of activation maps as done in Sect. 3.

We use the same GP language and CNN system (VGG-19) as in Sect. 3.1, but now we will use the CNN's top-level fully-connected (FC) layer. The output from the FC layer represents high-level features in the image data. The VGG-19 network is pre-trained on the ILSVRC-2012 dataset containing 1.3 million images denoting 1000 classes.[1] To use the FC layer, we provide fitness with a class ID number n of interest $(1 \leq n \leq 1000)$. Fitness then maximizes its class score $Class\ Score_n$ as read from the top-level layer. We do not minimize other classes for a couple of reasons. Firstly, doing so creates a difficult search space to navigate. Also, different classes related to the target class can naturally exhibit high activation values. It seems appropriate to encourage such behaviour, providing the target class is used in fitness evaluation.

While considering classes to use, we know that GP is unlikely to produce images for some of the more specialized classes, for example, those denoting exotic breeds of dogs. We therefore chose classes that seem fairly generic and obvious, for example, banana, orange, and others. All the experiments run for 25 generations, and we perform 8 runs per class.

4.2   Results

The results are shown in Fig. 6. Public-domain photographs showing visual examples of the target labels are given, along with their classification accuracy scores when given to the CNN. Note that these photographs are not necessarily part of the VGG training set.

---

[1] http://image-net.org/challenges/LSVRC/2012/.

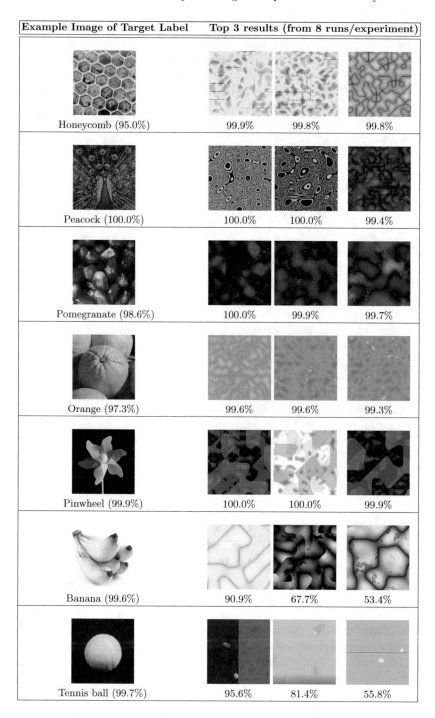

Fig. 6. Results from fully connected CNN experiments. One experiment per row. Classification accuracy per image is given (higher values preferred).

The first 5 target classes were easily mastered by GP, and the classification scores of many evolved images match or exceed the sample photograph scores. The solution images evoke aspects of the target class' colour, texture, and/or shape. For the peacock, for example, the feather plumage patterns are reproduced in the first 2 solutions, while the peacock's purple body colour also appears in the last image. The orange case is interesting, as the results do not show a spherical shape of an orange, but just capture the colour and possible texture. Pinwheel solutions show large curved areas of solid colours, similar to the pinwheel photograph.

Surprisingly, the banana and tennis ball runs were more challenging, despite what we expected to be easy subject matter. Banana solutions showed different degrees of yellow, tan and black, with low-frequency texture patterns. But the images are not strongly evocative of bananas. With the tennis ball images, the best results have green dots on larger solid or gradient backgrounds. We presume that the training images of tennis ball in the ILSVRC dataset have light-green tennis balls laying upon a flat tennis court. Interestingly, the best tennis ball solution has a small but clearly defined ball shape in the centre-bottom of the image. Since deep CNN classification is scale and position invariant, even a small ball shape is enough to generate a high classification score.

In summary, when using the fully connected CNN, results are highly dependent upon the nature of the pre-trained network. Specific image features seen in the training image set are used by the CNN for classification. The features encapsulated by the trained network must non-linearly differentiate the different classes of subject matter. This means that there is no consistent behaviour in the features used for different subjects. For example, a spherical shape is not required for an orange, but it is required for a tennis ball. Our results show that the fully-connected CNN is intriguing as a tool for evo-art. However, the current state of deep CNN technology also means that results may be unsatisfactory, should one expect the wholistic appearance of subject matter to arise in evolved images.

5   Conclusion

This paper presents a successful integration of a pre-trained, deep CNN within an evo-art system. We first showed how high-level activation nodes in the CNN can be exploited by fitness evaluation, in order to evolve images sharing characteristics of a target image. Our MMMS heuristic determined relevant activation nodes by examining activation levels over a family of images sharing characteristics of interest. This heuristic approach is necessary because of the complexity and opacity of deep neural networks, which are essentially black boxes with abstruse internal behaviours. Our results were promising for simple target images, as the evolved results often showed colours and shapes matching a target image. This is especially remarkable given that we did not train the VGG network, nor had access to the library of images used in its training. However, not all runs were successful. This is to be expected with evolutionary computation in many applications, depending on the performance of a particular run. In addition, although our texture language has a rich set of mathematical and noise functions, it has a natural bias towards the generation of particular styles of images, which might not necessarily conform to the target image.

We also used the complete fully-connected CNN, with its highest-level classification layer. These runs circumvent the need to examine the internal workings of the network, as done earlier with MMMS. Results are intriguing, especially from the insights they give regarding the differences between CNN and human classification. The training done on the VGG system has fine-tuned the network to accurately differentiate 1000 subject categories. To realize its impressive performance, however, the network distinguishes features which might seem incidental to a human viewer. For example, an orange is an orange texture, and a peacock is a green feather pattern. Such results may be of limited use to a user who wants an evolved image to show an orange's spherical shape, or an image of a bird. If we had trained the VGG network ourselves, we could have better controlled the forms of images evolved.

Our results with the fully-connected system confirm those of Nguyen *et al.* [21]. They used evolutionary computation to trick the classification abilities of deep CNNs. They evolved images with high classification scores, but which bore little resemblance to the subject matter being identified. This shows that visual CNN classification differs significantly from that done by humans. At least with our evolved images, it is often possible to intuit relationships between the evolved images and their target subjects.

There are many possible directions for future research. The bias of the procedural texture language used by the GP system can be leveraged, which will help to evolve images that better match particular styles of target images. More complex fitness strategies are possible, for example, multi-objective fitness might be used to separately score different activation matrices. The MMMS heuristic could be further refined. A more effective strategy for multiple target images should be considered.

In conclusion, as research advances and deep CNNs are better understood, they will be more effectively exploited in evo-art applications.

Acknowledgements. This research was supported by NSERC Discovery Grant RGPIN-2016-03653.

References

1. Dawkins, R.: The Blind Watchmaker. Norton & Company, Inc. (1986)
2. Sims, K.: Artificial evolution for computer graphics. In: Proceedings of the 18th Annual Conference on Computer Graphics and Interactive Techniques, vol. 25, no. 4, pp. 319–328, July 1991
3. Rooke, S.: Eons of genetically evolved algorithmic images. In: Bentley, P., Corne, D. (eds.) Creative Evolutionary Systems, pp. 339–365. Morgan Kaufmann, San Francisco (2002)
4. Todd, S., Latham, W.: Evolutionary Art and Computers. Academic Press, London (1992)

5. Bentley, P.: Creative Evolutionary Systems. Morgan Kaufmann, San Francisco (2002)
6. Romero, J., Machado, P.: The Art of Artificial Evolution. Springer, Heidelberg (2008). https://doi.org/10.1007/978-3-540-72877-1
7. Graf, J., Banzhaf, W.: Interactive evolution of images. In: Proceedings 4th Evolutionary Programming, pp. 53–65. MIT Press (1995)
8. den Heijer, E., Eiben, A.E.: Comparing aesthetic measures for evolutionary art. In: Di Chio, C., et al. (eds.) EvoApplications 2010. LNCS, vol. 6025, pp. 311–320. Springer, Heidelberg (2010). https://doi.org/10.1007/978-3-642-12242-2_32
9. Simonyan, K., Zisserman, A.: Very deep convolutional networks for large-scale image recognition. CoRR abs/1409.1556 (2014)
10. Donahue, J., et al.: DeCAF: a deep convolutional activation feature for generic visual recognition. In: International Conference on Machine Learning, pp. 647–655 (2014)
11. Krizhevsky, A., Sutskever, I., Hinton, G.: Imagenet classification with deep convolutional neural networks. In: Proceedings 25th International Conference on Neural Information Processing Systems, vol. 1, pp. 1097–1105. Curran Associates Inc. (2012)
12. Gatys, L., Ecker, A., Bethge, M.: Image style transfer using convolutional neural networks. In: Proceedings Computer Vision and Pattern Recognition, pp. 2414–2423. IEEE, June 2016
13. Lowe, D.: Object recognition from local scale-invariant features. In: Proceedings 7th IEEE International Conference on Computer Vision, vol. 2, pp. 1150–1157. IEEE (1999)
14. Dalal, N., Triggs, B.: Histograms of oriented gradients for human detection. In: Proceedings Computer Vision and Pattern Recognition, pp. 886–893. IEEE (2005)
15. Bay, H., Tuytelaars, T., Van Gool, L.: Speeded up robust features (SURF). Comput. Vis. Image Underst. **110**(3), 346–359 (2008)
16. Tanjil, F.: Deep learning concepts for evolutionary art. Master's thesis, Department Computer Science, Brock U. (2018)
17. Koza, J.: Genetic Programming: On the Programming of Computers by Means of Natural Selection. MIT Press, Cambridge (1992)
18. Poli, R., Langdon, W., McPhee, N.: A Field Guide to Genetic Programming. Lulu Enterprises UK Ltd. (2008)
19. Goodfellow, I., Bengio, Y., Courville, A.: Deep Learning. MIT Press, Cambridge (2016)
20. Gooch, B., Gooch, A.: Non-photorealistic Rendering. A. K. Peters (2001)
21. Nguyen, A., Yosinski, J., Clune, J.: Deep neural networks are easily fooled: high confidence predictions for unrecognizable images. In: 2015 IEEE Conference on Computer Vision and Pattern Recognition (CVPR), pp. 427–436, June 2015
22. Bontrager, P., Lin, W., Togelius, J., Risi, S.: Deep interactive evolution. In: Liapis, A., Romero Cardalda, J.J., Ekárt, A. (eds.) EvoMUSART 2018. LNCS, vol. 10783, pp. 267–282. Springer, Cham (2018). https://doi.org/10.1007/978-3-319-77583-8_18
23. Stanley, K.: Compositional pattern producing networks: a novel abstraction of development. Genet. Program. Evolvable Mach. **8**(2), 131–162 (2007)
24. Baluja, S., Pomerleau, D., Jochem, T.: Towards automated artificial evolution for computer-generated images. Connection Sci. **6**(2–3), 325–354 (1994)
25. Agapitos, A., et al.: Deep evolution of image representations for handwritten digit recognition. In: Proceedings CEC 2015, Sendai, Japan, 25–28 May 2015, pp. 2452–2459. IEEE (2015)

26. Gircys, M.: Image evolution using 2D power spectra. Master's thesis, Brock University, Department of Computer Science (2018)
27. Luke, S.: ECJ: a Java-based evolutionary computation research system. https:// cs.gmu.edu/~eclab/projects/ecj/. Accessed 16 Sept 2017
28. Chintala, S.: Pytorch documentation. http://pytorch.org/docs/master/. Accessed 16 Sept 2017
29. Chintala, S.: PyTorch: tensors and dynamic neural networks in python with strong GPU acceleration. http://pytorch.org/. Accessed 16 Sept 2017

Adversarial Evolution and Deep Learning – How Does an Artist Play with Our Visual System?

Alan Blair[✉]

School of Computer Science and Engineering,
University of New South Wales, Sydney, Australia
blair@cse.unsw.edu.au

Abstract. We create artworks using adversarial coevolution between a genetic program (HERCL) generator and a deep convolutional neural network (LeNet) critic. The resulting artificially intelligent artist, whimsically named Hercule LeNet, aims to produce images of low algorithmic complexity which nevertheless resemble a set of real photographs well enough to fool an adversarially trained deep learning critic modeled on the human visual system. Although it is not exposed to any pre-existing art, or asked to mimic the style of any human artist, nevertheless it discovers for itself many of the stylistic features associated with influential art movements of the 19th and 20th Century. A detailed analysis of its work can help us to better understand the way an artist plays with the human visual system to produce aesthetically appealing images.

Keywords: Evolutionary art · AI-generated art ·
Artist-critic coevolution · Adversarial training ·
Computational creativity

1 Introduction

There has recently been renewed interest in the paradigm of artist-critic coevolution or adversarial training in which an artist (generator) tries to produce images which are similar in some way to a set of real images, and a critic tries to discriminate between the real images and those generated by the artist [1,2].

The earliest work in this area followed an interactive evolution scenario, with a human playing the role of the critic, and the artist trained by some form of evolutionary computation such as Biomorphs [3], Genetic Programming [4], Cellular Automata [5] or Compositional Pattern Producing Networks [6]. In these systems, several candidate images appear on the screen and the user is invited to select one or more of them for inclusion in the next generation. These approaches have produced some remarkable images, but the process can be time-consuming for the human as several dozen generations are often required in order to produce a pleasing image.

© Springer Nature Switzerland AG 2019
A. Ekárt et al. (Eds.): EvoMUSART 2019, LNCS 11453, pp. 18–34, 2019.
https://doi.org/10.1007/978-3-030-16667-0_2

To save human effort, the critic has sometimes been replaced with a pre-trained image or face detection system [7,8], or a Convolutional Neural Network (CNN) trained to mimic human preferences [9]. Similar machine learning approaches have been used to explore aesthetic feature selection [10,11] or to model collective artistic behavior using Self Organizing Maps [12].

An exciting new paradigm was introduced in 2008 where images are created through an adversarial arms race between a creator (generator) and an adaptive critic [13]. The critic is rewarded for its ability to distinguish "real" images from those generated by the creator, while the creator is rewarded for generating images that will fool the critic into thinking they are real. Typically, a genetic algorithm was used for the creator, while the critic was a 2-layer neural network trained to classify images based on certain statistical features extracted from the image [14–16].

A powerful new variant of this adversarial paradigm was developed in 2014 known as Generative Adversarial Networks (GANs) [17], where the generator and discriminator (critic) are both CNNs trained by gradient descent. Compared to other approaches, GANs produce astonishingly realistic images [18], and also have the advantage that a single trained network can produce a great variety of images from latent variables.

The aim of the present work is not to generate photo-realistic images, nor to copy existing artistic styles, but rather to create new abstract artworks sui generis. Our system employs a recently introduced hybrid approach [19] where the generator is a Genetic Program evolved by hierarchical evolutionary recombination (HERCL) [20], while the critic is a GAN-style convolutional neural network (LeNet) [21] trained by gradient descent. Each image is generated via a genetic program which takes as input the x and y coordinates of a pixel, and produces as output the R, G, B values assigned to that pixel in the image [22].

Because it is comprised of a HERCL generator and a LeNet critic, the resulting adversarial art generation system is sometimes whimsically referred to as if it were a real artist named "Hercule LeNet".

2 Interplay Between Generator and Critic

It is important to note that this hybrid evolution and deep learning system is not trained on any pre-existing artworks, nor is it asked to mimic the style of any human artist, nor to directly satisfy the whims of any human observer. Rather, it is presented with a number of real images, and aims to produce synthetic images which have low algorithmic complexity (due to selective pressure for shorter programs), but which nevertheless resemble the real images well enough to fool an adversarially trained (LeNet) critic with an architecture modeled on the human visual system.

This very broad objective effectively liberates the artificial artist from stylistic preconceptions, allowing it to freely explore and develop its own artistic style(s), which can then be analysed.

In the context of image classification, it has been pointed out in a number of contexts that deep convolutional networks are easily fooled [8]. This is certainly a problem if accurate recognition is the goal, or if the networks are fooled in ways which appear puzzling to a human observer. However, if the networks are fooled in ways which seem quite familiar to us as humans, this might indicate that the networks share certain quirks of the human visual system – quirks which can be exploited in interesting ways by abstract visual artists.

When it comes to impressionist, minimalist or abstract art, we as humans are not actually deceived into thinking that the images are real; rather, our appreciation of these artworks relies at least in part on their ability to challenge and tantalize our visual system, prompting us to look at things from a new perspective. If our evolutionary artist (HERCL) is able to fool the artificial visual system of the convolutional network (LeNet) in an analogous manner, it may produce images which have a similar appeal for human observers, and help us to gain a better understanding of visual aesthetics.

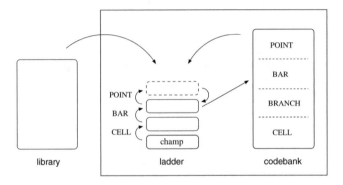

Fig. 1. Hierarchical evolutionary re-combination. If the top agent on the ladder becomes fitter than the one below it, the top agent will move down to replace the lower agent (which is transferred to the codebank). If the top agent exceeds its maximum number of allowable offspring without ever becoming fitter than the one below it, the top agent is removed from the ladder (and transferred to the codebank). Code from related tasks (in this case, the code that generated earlier images in the gallery) is kept in a library and made available for genetic recombination.

3   Adversarial Training Paradigm

Our adversarial training paradigm, essentially the same as in [19], is as follows: In the first round, the LeNet critic is trained to assign a low cost (close to 0) to all the real images and a high cost (close to 1) to a blank (i.e. completely white) image. Training is by stochastic gradient descent using the Adam optimizer [23]. In all subsequent rounds, a HERCL artist is evolved with the aim of producing a new image to which the current critic will assign as low a cost as possible — continuing until the cost becomes lower than 0.01, or a maximum of 200000 images have been generated. The resulting minimal-cost image is then added to a "gallery" of images, and a new critic is trained to assign a low cost to the real images and a high cost to all the images in the gallery (i.e. the minimal-cost image produced by the artist in every previous round). This process continues for a fixed number of rounds (typically, between 600 and 1000) with every round adding exactly one new image to the gallery.

Table 1. HERCL commands

Input and Output		Stack Manipulation and Arithmetic	
i	fetch INPUT to input buffer	#	PUSH new item to stack ...... $\mapsto$ x
s	SCAN item from input buffer to stack	!	POP top item from stack ......$x \mapsto$ ......
w	WRITE from stack to output buffer	c	COPY top item on stack ......$x \mapsto$ ......x, x
o	flush OUTPUT buffer	x	SWAP top two items ...$y, x \mapsto$ ...x, y
Registers and Memory		y	ROTATE top three items $z, y, x \mapsto x, z, y$
		−	NEGATE top item ......$x \mapsto$ .....$(-x)$
<	GET value from register	+	ADD top two items ...$y, x \mapsto$ ...$(y+x)$
>	PUT value into register	*	MULTIPLY top two items ...$y, x \mapsto$ ...$(y * x)$
^	INCREMENT register	**Mathematical Functions**	
v	DECREMENT register		
{	LOAD from memory location	r	RECIPROCAL ..$x \to$ ..$1/x$
}	STORE to memory location	q	SQUARE ROOT ..$x \to$ ..$\sqrt{x}$
Jump, Test, Branch and Logic		e	EXPONENTIAL ..$x \mapsto$ ..e^x
		n	(natural) LOGARITHM ..$x \mapsto$ ..$\log_e(x)$
j	JUMP to specified cell (subroutine)	a	ARCSINE ..$x \mapsto$ ..$\sin^{-1}(x)$
|	BAR line (RETURN on . | HALT on 8 |)	h	TANH ..$x \mapsto$ ..$\tanh(x)$
=	register is EQUAL to top of stack	z	ROUND to nearest integer
g	register GREATER than top of stack	?	push RANDOM value to stack
:	if TRUE, branch FORWARD	**Double-Item Functions**	
;	if TRUE, branch BACK		
&	logical AND	%	DIVIDE/MODULO..$y, x \mapsto$ ..$(y/x), (y \bmod x)$
/	logical OR	t	TRIG functions ..$\theta, r \mapsto$ ..$r \sin \theta, r \cos \theta$
~	logical NOT	p	POLAR coords..$y, x \mapsto$ ..atan2$(y,x), \sqrt{x^2+y^2}$

Within each round, a new artist is evolved using hierarchical evolutionary re-combination, as illustrated in Fig. 1 (see [19,20] for further details). HERCL is a general-purpose evolutionary computation paradigm which has previously been applied to tasks such as classification [24], control [25], string processing [26] and line drawing [27]. It uses a simple imperative language with instructions for manipulating a stack, registers and memory, thus combining features from linear GP and stack-based GP. The full list of HERCL commands is given in Table 1. During evolution, HERCL code from previous or related tasks (in this case, the code that produced every previous image in the gallery) is kept in a library and made available as material for genetic recombination, so that the artist has the opportunity to explore variations on an earlier theme.

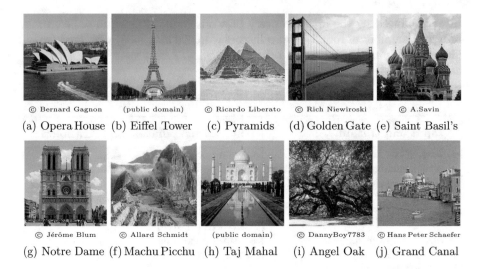

© Bernard Gagnon   (public domain)   © Ricardo Liberato   © Rich Niewiroski   © A.Savin

(a) Opera House   (b) Eiffel Tower   (c) Pyramids   (d) Golden Gate   (e) Saint Basil's

© Jérôme Blum   © Allard Schmidt   (public domain)   © DannyBoy7783   © Hans Peter Schaefer

(g) Notre Dame (f) Machu Picchu (h) Taj Mahal   (i) Angel Oak   (j) Grand Canal

Fig. 2. Examples of real images for each of the 10 landmarks.

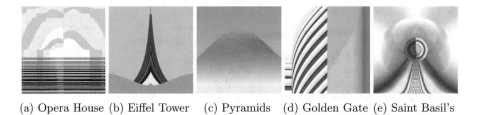

(a) Opera House (b) Eiffel Tower   (c) Pyramids   (d) Golden Gate   (e) Saint Basil's

Fig. 3. Generated images from preliminary experiments using small critic network and no data augmentation.

Note that the only role played by the human in this process is in searching for and selecting images of the desired subject, and scanning through the 600 to 1000 images produced, to select a subset of images that are considered to have the greatest visual appeal. Thus, the human acts purely as an initial patron and final curator of the artworks; the process of actually generating them is completely autonomous and untouched by human hands.

4   Preliminary Experiments and Enhancements

For most of the experiments presented in [19], the real images covered the full range of CIFAR-10 categories [28], including planes, cars, birds, cats, deer, dogs, frogs, horses, ships and trucks. The generated images were intriguing, and exhibited a broad range of artistic styles; but, because the real images were so heterogenous, the generated images lacked a sense of coherence, and no image could really be attributed to, or said to "depict", any particular subject.

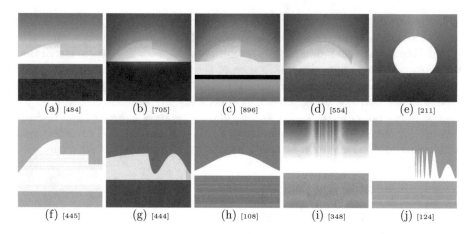

Fig. 4. Sydney Opera House

We therefore made a plan to extend that work with a new set of experiments in which photographs of a famous landmark would constitute the real images for each experimental run. We chose 10 landmarks, and in each case performed a Google image search and selected between 8 and 45 images which we felt would best serve as real images for that landmark (See Fig. 2).

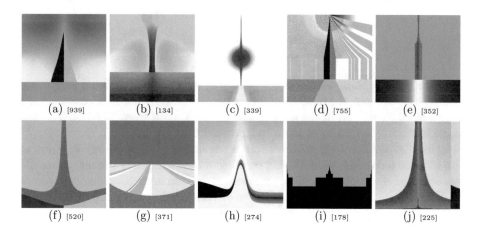

Fig. 5. Eiffel Tower

We first ran a set of five preliminary experiments, using the landmarks from Fig. 2(a) to (e), and the same network structure used in [19], namely: a convolutional layer with 6 filters (5 × 5), a max pooling layer with stride 2, a second convolutional layer with 16 filters (5 ×5), another max pooling layer with stride 2, two fully connected layers of size 120 and 84, and two output units. Leaky ReLUs were used at all the hidden nodes, and softmax at the output. The real images were automatically converted to a resolution of 48×48, and all images generated by the artist were also rendered at a resolution of 48×48. The single image that we selected as most appealing from each of these five experiments are collected in Fig. 3. Although these images do have some artistic merit, we felt that the system could perhaps be improved by increasing the number and size of the convolutional filters in the critic, and by employing data augmentation to compensate for the paucity of training images [29]. We also wanted to experiment with changing the resolution at which images were presented to the critic.

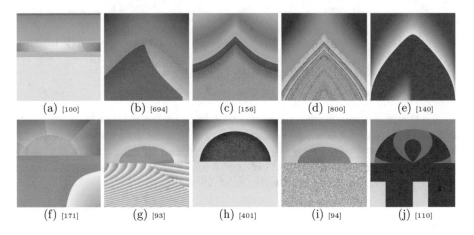

(a) [100]   (b) [694]   (c) [156]   (d) [800]   (e) [140]

(f) [171]   (g) [93]   (h) [401]   (i) [94]   (j) [110]

Fig. 6. Great Pyramid

With this in mind, we ran two new sets of experiments, denoted Res48 and Res64. In both cases, the number of filters was increased in the first convolutional layer from 6 to 16, and in the second convolutional layer from 16 to 24. The size of the filters in the first layer was also increased from 5×5 to 7×7. For the Res48 experiments, images were generated at a resolution of 50×50 but cropped to 48×48 when fed to the critic. During training of the critic, data augmentation [29] was achieved by cropping both the real and generated images randomly in one of 9 different ways (stripping 0, 1 or 2 rows or columns from one end of the image and a complementary number from the other end). During evolution of the artist, exactly 1 row or column was stripped from all sides of the image and only the central 48×48 region was fed to the critic. For the Res64 experiments, images were generated at a resolution of 68×68 and cropped to 64×64 in one of 25 different ways for training of the critic, with exactly 2 rows and columns stripped from all sides of the image during evolution of the artist. The experiments were run until they had produced 1000 images for Res48, or 600 images for Res64 (in each case, about two weeks of computation on a single CPU).

5   Results, Artistic Styles and Analysis

Five selected images from each experiment are shown in Figs. 4, 5, 6, 7, 8, 9, 10, 11, 12 and 13.[1] In each case, Subfigures (a) to (e) are from Res48 and (f) to (j) from Res64. The small numbers in square brackets indicate the round at which the image was generated.

5.1   Minimalism and Abstraction

Some of the generated images resemble minimalist art, where the subject is suggested with simple patters of colors and shapes. These include Figs. 4(a, e, f, g), 5(a, f), 6(a, f), 7(a, b, f), 8(a, b), 9(f), 10(a, e, f, g, j), 11(a, g), 12(f, g) and 13(f). Subtle yet important shading effects may be added to create an illusion of depth, as in Figs. 5(j), 6(b, g, i), 7(c), 10(b, c, d, h), 12(i) and 13(b), or even to add a luminous quality, as in Figs. 4(b, c, d) and 6(h).

Convolutional neural networks with max-pooling layers and (Leaky) Rectified Linear Units generally make their classification based on an accumulation of visual features supporting the hypothesis of realness or fakeness. This semi-linear aggregation of evidence often enables the artist to fool the critic by providing some visual features in abundance while omitting others. This is evident in Figs. 4(i), 5(g), 6(a) and 11(c) where the landscape is rendered but the building itself is missing; in Fig. 7(a, h, i) the vertical beams of the bridge are rendered but not its overall shape; Fig. 11(e) includes the circular feature at the center of Notre Dame, and the general circular motif, but other features are absent. Figure 13(h) captures the color and texture (concentrated in the vertical direction) of buildings along the Grand Canal in Venice, but not their outline.

Components of the subject may be manipulated, such as in Fig. 11(d) where the reflecting pool is visible but tilted at a weird angle, or Fig. 5(b) which gives the impression that a mischievous giant has uprooted the tower and stuck it back in the ground upside-down. Figures 8(a) and 9(a, j) have a slightly art deco look about them, whilst others such as Figs. 6(j), 8(j) and 13(a, e) seem to bear little resemblance to the underlying object, but instead make playful use of simple geometric forms in a limited range of colors, in the spirit of suprematism.

Figure 11(e) looks rather like a bird, totem pole or kite; yet, when we compare it to a real image of the Taj Mahal in Fig. 2(h), we see that the system has indeed extracted certain regularities of shape and color — but they are distinct from those that a human would normally focus on, thereby giving the work an abstract rather than figurative quality.

5.2   Colors and Shading

Art movements such as impressionism and fauvism rely in part on the propensity of the human visual system to respond to relative color rather than absolute color. This is often also the case for convolutional neural networks, because the

---

[1] © Hercule LeNet (all images in Figs. 3, 4, 5, 6, 7, 8, 9, 10, 11, 12 and 13).

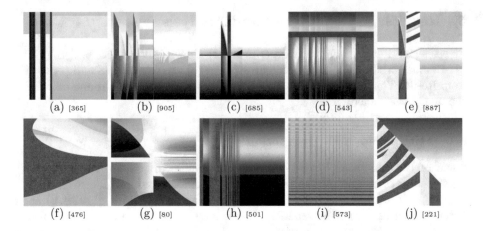

(a) [365]   (b) [905]   (c) [685]   (d) [543]   (e) [887]

(f) [476]   (g) [80]   (h) [501]   (i) [573]   (j) [221]

Fig. 7. Golden Gate Bridge

early convolutional layers can learn to respond to differences in intensity among pixels in a local neighborhood, rather than absolute intensity. Our HERCL artist can exploit this feature of the LeNet critic by altering the colors of familiar objects. The Great Pyramid or the Eiffel Tower may become red; the Taj Mahal or Notre Dame may take on all the colors of the rainbow. In Figs. 4(b, c, d), 5(a), 6(b, c, d, e, g, h, i) and 11(i) a halo effect is created to heighten the contrast (in one or more color channels) between the building and the sky. Some images such as Figs. 6(h), 8(b), 10(g), 11(j) and 13(j) are rendered in vibrant, perhaps even fauvist colors; others such as Figs. 7(c) and 10(h) are almost black-and-white in nature, but catch our attention through contrast and shading. Our HERCL artist seems to show a preference for vibrant reds and greens, which may be due in part to the encoding of the image in R,G,B space corresponding to the primary colors for mixing paints as well as the red, green and blue receptors of the human retina. It would be interesting in future work to change the encoding to Y,U,V space and see whether a different palette of colors would then be preferred by the artist.

There are also cases where the artist has taken care to accurately reproduce the natural color. This typically occurs in situations where sections of the original photographs consist of a featureless expanse—such as the desert sands, or the sky over the Great Pyramid, Notre Dame or the Sydney Opera House.

Figure 8(b) provides an abstract rendition of Saint Basil's Cathedral in the form of colored ribbons. Comparing it to a real image in Fig. 2(e), we see that it mimics both the shapes and colors of the original subject, but in new combinations. A central dome with a spire (depicted in red) appears to be flanked by two smaller, striped domes, with a couple of wispy white clouds, and black corners for the surrounding trees; blue, pink and yellow ribbons capture the colors of the blue dome, reddish towers and light colored base, while the green stripes reflect the green of the statue, the gables and the small tented roof.

5.3   Fractals and Texture

The selective pressure for images of low algorithmic complexity often entices the artist to employ self-similarity or fractals. Figures 8(c, d), 9(c, d, i), 11(d, j), 12(a, b, c, d, e) and 13(c, i) are recognizable as fractal art, and also exhibit some of the qualities of psychedelic art.

Figure 7(i) uses fractal geometry to tease the mind of the observer in a manner similar to the works of M.C. Escher. When we look at this image (best viewed in color) it is obvious that the red beams are reflected in the water; yet, we cannot quite discern where the sky meets the sea, or where the real beam ends and the reflection begins. Thus, the eye of the observer is led around in endless circles, searching for elusive boundaries that never quite resolve themselves.

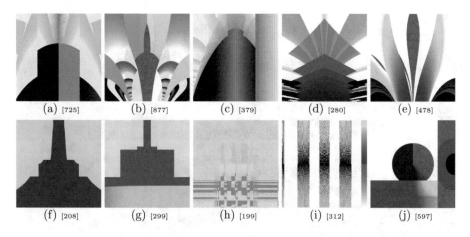

Fig. 8. Saint Basil's Cathedral (Sobor Vasiliya)

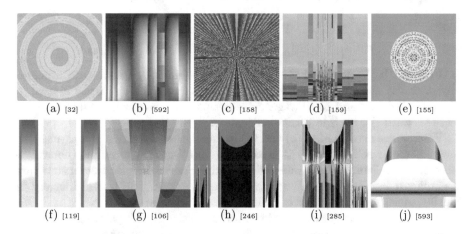

Fig. 9. Notre Dame de Paris

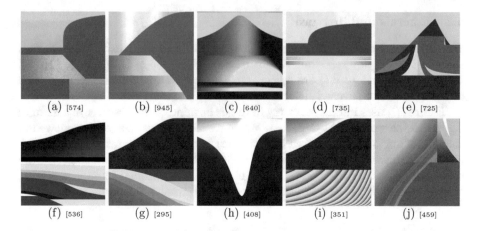

Fig. 10. Machu Picchu

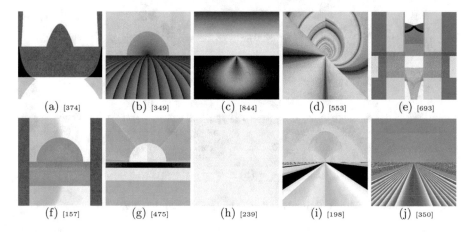

Fig. 11. Taj Mahal

Figures 4(h, j) and 10(d) use horizontal self-similarity to simulate topography or waves and ripples on the water, while vertical self-similarity is used in Figs. 4(i), 7(d, h, i) and 13(h) to generate beams or reeds growing thinner as they recede toward the horizon (or its vertical equivalent).

Fractals are used in Fig. 8(c) and (d) to create the impression of a psychedelic tower. Circular self-similarity is used in Figs. 9(a) and (e) to mimic the centerpiece of Notre Dame Cathedral, while a distinctive rectangular fractal structure is employed in Fig. 9(d) to simulate its overall shape. The fractal imagery in Fig. 9(i) gives the appearance of stained glass or glass art.

All the fractal images can be blown up to the size of wall paintings, with fine details emerging at the new scale (such as the delicate structure of the wings in Fig. 12(e)). In parts of the images where the function defining the R,G,B values

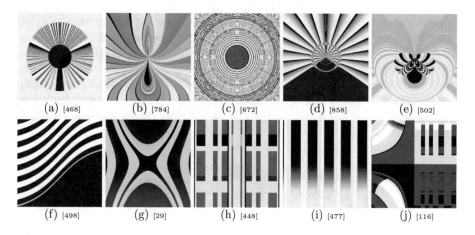

(a) [468]   (b) [784]   (c) [672]   (d) [858]   (e) [502]

(f) [498]   (g) [29]   (h) [448]   (i) [477]   (j) [116]

Fig. 12. Angel Oak Tree

becomes extremely sensitive to changes in x and y coordinates, a seemingly random dot pattern can emerge, creating a form of pointillism. This phenomenon produces an exotic texture in Fig. 8(i) and a "stardust" effect in Fig. 5(d). In Fig. 6(d) a pattern that is regular near the outside of the pyramid appears to gradually break down to a random collection of dots as we move toward the center, thus achieving a kind of sandy texture. When Figs. 6(d) and (i) are enlarged, we have the impression of looking out over a vast desert within which we can see individual grains of sand.

Figure 13(d) may at first appear to be a formless smattering of dots. However, with some imagination, we can think of it as conveying the general impression of a landscape reflected in the water, with a ridge along the waterline, some kind of indiscernible structures on the right, and open space to the left.

There is a tradition, dating back to Kazimir Malevich in 1915, of exhibiting an ostensibly blank canvas which contains no representational content in the usual sense, but instead invites the viewer to appreciate the texture of the paint or the detail of the brush strokes. Very occasionally, a similar phenomenon can be observed in our system; Fig. 11(h) at first appears blank, but on closer inspection we can see that it contains fine details of texture. The LeNet critic, trained to reject 238 previous images in the gallery, nevertheless assigns a probability of only 30% for this image to be synthetic.

5.4 Metaphor

There are several instances where images have been evolved to resemble one thing, but end up reminding us of something else. The Opera House becomes a rising moon in Fig. 4(e); the Eiffel Tower looks like a cardboard cutout in Fig. 5(j); the Golden Gate Bridge can inspire a sailing rig in Fig. 7(e) or a collection of colored scarves hanging on a line in Fig. 7(j). Saint Basil's Cathedral prompts an image of a flower in Fig. 8(e); Notre Dame elicits an image resembling either a bed or a motor car in Fig. 9(j); the branches of an oak tree are morphed into a psychedelic insect in Fig. 12(e). Of course, our automated system has no knowledge of insects, ships, scarves, beds or flowers. It is possible that these metaphors emerge simply by chance. However, they may also reflect certain universal similarities that exist between various types of objects. If we suppose that Saint Basil's Cathedral was originally designed to resemble a flower, it may be that the wily artist has somehow brought to life the original conception behind the artifice.

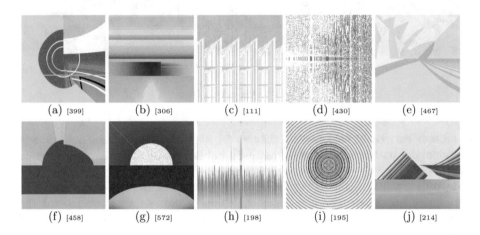

(a) [399]   (b) [306]   (c) [111]   (d) [430]   (e) [467]

(f) [458]   (g) [572]   (h) [198]   (i) [195]   (j) [214]

Fig. 13. Grand Canal in Venice

One of the most intriguing metaphors appears in Fig. 13(j) where a row of buildings along the Grand Canal in Venice have been rendered to look like a sailing ship. If we look closely at one of the original images in Fig. 2(j) we can start to understand how the artist may have "mistaken" a triangular roof in the middle of the image for the prow of a ship, the buildings to its right for the hull, and the sloping roofs above for the sails. This example neatly illustrates the ability of the adversarial artist/critic system to surprise us by discovering connections and similarities that may otherwise have escaped our notice.

5.5 Repeated Substructures with Variations

In many cases we see substructures repeated within an image, but with minor variations. These include Figs. 5(g), 7(j), 8(b, d, e), 9(b, d, g, i) and 12(b, h, j). As noted in [27,30,31] these imperfectly repeated substructures give the impression of having arisen organically through some natural process. In Fig. 13(j) the formless mass on the left appears to have been broken down and reconfigured into the sailing ship on the right; if we were to take the liberty of anthropomorphising the artist, we might be tempted to say that this work symbolises the use of raw materials to construct a fabricated object.

5.6   Genotype to Phenotype Mapping

The HERCL code for the evolved images ranges in length from 10 to 223 characters, with a median of 73 characters (by convention, we treat the boundary between subroutines as a single character). The code length for the 100 selected images ranges from 36 to 162, with a median of 79 characters. As an example, the code (of length 73) for generating Fig. 13(j) is shown in Table 2, along with equivalent pseudocode. Note that x runs from left (-1) to right $(+1)$, while y runs from top (-1) to bottom $(+1)$. Color intensities are output in the order blue, green, red; outputs less than 0 or greater than 1 are treated as 0 and 1, respectively. In order to avoid floating point exceptions, HERCL adheres to certain safety conventions, such as $\sin^{-1}(\alpha) = -\pi/2$ for $\alpha < -1$, which in this case ensures a uniform color for the sky.

Table 2. HERCL code and Pseudocode for generating Fig. 13(j)

HERCL code:

```
0[!qatcz]
1[capwwwo.]
2[%]
3[is.32#>sg:1j|c>xg:hp2j|+a{>cpa%.4338#p>g~<:0j|xww.88#wo]
```

Pseudocode:

   scan (x, y) // $-1 \le x \le 1$,   (upper) $-1 \le y \le 1$ (lower)

   if $y \ge 0.32$ // water

      return $\left(\sqrt{y^2 + (\sin^{-1} y)^2},\ \text{atan2}(y, \sin^{-1} y),\ x\right)$

   else

      if $y > x$ // obstacle

         $u = \sin^{-1}(x + y)$

      else   // ship

         $r = \sqrt{y^2 + \tanh(x)^2},\quad \theta = \text{atan2}(y, \tanh(x))$

         $u = \sin^{-1}(\lfloor \theta/r \rfloor + (\theta \bmod r))$

      end

      $\phi = \frac{\pi}{4}(-1 + 2\,\text{sgn}(u)),\quad \rho = \sin^{-1}(\sqrt{2}u)$

      $z = \text{atan2}((\phi \bmod \rho), 0.4338),\quad s = \sqrt{(\phi \bmod \rho)^2 + 0.4338^2}$

      if $s \le z$ // sails

         return $(z, s, 0.88)$

      else   // hull

         $v = \sin^{-1}(\sqrt{z})\cos(\lfloor \phi/\rho \rfloor)$

         return $(v, \lfloor v \rfloor, 0.88)$ // (blue, green, red)

      end

   end

6   Discussion

In general terms, the artistic quality across all landmarks (Figs. 4, 5, 6, 7, 8, 9, 10, 11, 12 and 13) of the images from the Res48 experiments (Subfigures (a) to (e)) seems to be roughly comparable with those from Res64 (Subfigures (f) to (j)). Subjectively, if we were asked to select the single most appealing image from each landmark, our choice would probably be: Figs. 4(c), 5(d), 6(d), 7(j), 8(b), 9(d), 10(e), 11(e), 12(e) and 13(j)—thus, 8 images from Res48 and only 2 from Res64. On the other hand, of these 8 images from Res48, 6 of them were generated later than the 600th round, which was the cutoff point for Res64. So, it would appear that Res48 is more likely to produce a high quality artwork within the computational budget of two weeks of CPU time; but, we cannot exclude the possibility that Res64 might produce superior images if it were run for longer than 600 rounds.

7   Conclusion

We have demonstrated the ability of an autonomous artist to create abstract art based on famous landmarks. The art is not constrained by human input, but instead emerges naturally from a tradeoff between the selective pressure for low algorithmic complexity and the imperative to fool the critic — thus giving the artist freedom to create images in its own intrinsic style. Nevertheless, we can recognize certain fundamental elements associated with influential art movements of the 19th and 20th Century including minimalism, impressionism, fauvism, pointillism and suprematism as well as psychedelic and fractal art.

The pressure for low algorithmic complexity often leads to self-similarity, fractals or repeated substructures which are pleasing to the eye. Our appreciation of these works suggests there may be some component of the human visual system which specifically responds to self-similarity. In future work, we hope to explore how such a mechanism might be modeled artificially, and integrated into both generative and discriminative convolutional neural networks.

Acknowledgment. Thanks to Jacob Soderlund, Darwin Vickers and Tanakrit Udom-choksakul for contributing code, and to Jeff Clune, Ken Stanley, Yoshua Bengio, Oliver Bown and Gary Greenfield for fruitful discussions. This research was undertaken with the support of Akin.com, as well as resources from the National Computational Infras-tructure (NCI), which is supported by the Australian Government.

References

1. Galanter, P.: Computational aesthetic evaluation: past and future. In: McCormack, P., d'Inverno, M. (eds.) Computers and Creativity. Springer, Heidelberg (2012). https://doi.org/10.1007/978-3-642-31727-9_10
2. McCormack, J., Bown, O., Dorin, A., McCabe, J., Monro, G., Whitelaw, M.: Ten questions concerning generative computer art. Leonardo **47**(2), 135–141 (2014)

3. Dawkins, R.: The Blind Watchmaker: Why the Evidence of Evolution Reveals a World Without Design. Norton, New York (1986)
4. Sims, K.: Artificial evolution for computer graphics. ACM Comput. Graph. **25**(4), 319–328 (1991)
5. Kowaliw, T., Dorin, A., McCormack, J.: Promoting creative design in interactive evolutionary computation. IEEE Trans. Evol. Comput. **16**(4), 523–536 (2012)
6. Secretan, J., et al.: Picbreeder: a case study in collaborative evolutionary exploration of design space. Evol. Comput. **19**(3), 373–403 (2011)
7. Machado, P., Correia, J., Romero, J.: Expression-based evolution of faces. In: Machado, P., Romero, J., Carballal, A. (eds.) EvoMUSART 2012. LNCS, vol. 7247, pp. 187–198. Springer, Heidelberg (2012). https://doi.org/10.1007/978-3-642-29142-5_17
8. Nguyen, A., Yosinski, J., Clune, J.: Deep neural networks are easily fooled: high confidence predictions for unrecognizable images. In: IEEE Conference on Computer Vision and Pattern Recognition (CVPR), pp. 427–436 (2015)
9. Baluja, S., Pomerlau, D., Todd, J.: Towards automated artificial evolution for computer-generated images. Connection Sci. **6**(2), 325–354 (1994)
10. Ekárt, A., Sharma, D., Chalakov, S.: Modelling human preference in evolutionary art. In: Di Chio, C., et al. (eds.) EvoApplications 2011. LNCS, vol. 6625, pp. 303–312. Springer, Heidelberg (2011). https://doi.org/10.1007/978-3-642-20520-0_31
11. Correia, J., Machado, P., Romero, J., Carballal, A.: Feature selection and novelty in computational aesthetics. In: Machado, P., McDermott, J., Carballal, A. (eds.) EvoMUSART 2013. LNCS, vol. 7834, pp. 133–144. Springer, Heidelberg (2013). https://doi.org/10.1007/978-3-642-36955-1_12
12. Saunders, R., Gero, J.S.: Artificial creativity: a synthetic approach to the study of creative behaviour. In: Computational and Cognitive Models of Creative Design V, pp. 113–139. Key Centre of Design Computing and Cognition, University of Sydney (2001)
13. Machado, P., Romero, J., Manaris, B.: Experiments in computational aesthetics: an iterative approach to stylistic change in evolutionary art. In: Romero, J., Machado, P. (eds.) The Art of Artificial Evolution. Natural Computing Series, pp. 318–415. Springer, Heidelberg (2008). https://doi.org/10.1007/978-3-540-72877-1_18
14. Greenfield, G., Machado, P.: Simulating artist and critic dynamics – an agent-based application of an evolutionary art system. In: Proceedings of the International Joint Conference on Computational Intelligence (IJCCI), Funchal, Madeira, Portugal, pp. 190–197 (2009)
15. Li, Y., Hu, C.-J.: Aesthetic learning in an interactive evolutionary art system. In: Di Chio, C., et al. (eds.) EvoApplications 2010. LNCS, vol. 6025, pp. 301–310. Springer, Heidelberg (2010). https://doi.org/10.1007/978-3-642-12242-2_31
16. Correia, J., Machado, P., Romero, J., Carballal, A.: Evolving figurative images using expression-based evolutionary art. In: 4th International Conference on Computational Creativity (ICCC), pp. 24–31 (2013)
17. Goodfellow, I., et al.: Generative adversarial nets. In: Advances in Neural Information Processing Systems, pp. 2672–2680 (2014)
18. Nguyen, A., Clune, J., Bengio, Y., Dosovitskiy, A., Yosinski, J.: Plug & play generative networks: conditional iterative generation of images in latent space. In: IEEE Conference on Computer Vision and Pattern Recognition (CVPR), pp. 3510–3520 (2017)
19. Soderlund, J., Blair, A.: Adversarial image generation using evolution and deep learning. In: IEEE Congress on Evolutionary Computation (2018)

20. Blair, A.: Learning the Caesar and Vigenere Cipher by hierarchical evolutionary re-combination. In: IEEE Congress on Evolutionary Computation, pp. 605–612 (2013)
21. LeCun, Y., Bottou, L., Bengio, Y., Haffner, P.: Gradient-based learning applied to document recognition. Proc. IEEE **86**(11), 2278–2323 (1998)
22. Rooke, S.: Eons of genetically evolved algorithmic images. In: Bentley, P.J., Corne, D.W. (eds.) Creative Evolutionary Systems, pp. 339–365. Morgan Kauffmann, San Francisco (2002)
23. Kingma, D.P., Ba, J.L.: Adam: a method for stochastic optimization. In: International Conference on Learning Representations (ICLR), pp. 1–15 (2015)
24. Blair, A.D.: Transgenic evolution for classification tasks with HERCL. In: Chalup, S.K., Blair, A.D., Randall, M. (eds.) ACALCI 2015. LNCS (LNAI), vol. 8955, pp. 185–195. Springer, Cham (2015). https://doi.org/10.1007/978-3-319-14803-8_15
25. Blair, A.: Incremental evolution of HERCL programs for robust control. In: Genetic and Evolutionary Computation Conference (GECCO) Companion, pp. 27–28 (2014)
26. Soderlund, J., Vickers, D., Blair, A.: Parallel hierarchical evolution of string library functions. In: Handl, J., Hart, E., Lewis, P.R., López-Ibáñez, M., Ochoa, G., Paechter, B. (eds.) PPSN 2016. LNCS, vol. 9921, pp. 281–291. Springer, Cham (2016). https://doi.org/10.1007/978-3-319-45823-6_26
27. Vickers, D., Soderlund, J., Blair, A.: Co-evolving line drawings with hierarchical evolution. In: Wagner, M., Li, X., Hendtlass, T. (eds.) ACALCI 2017. LNCS (LNAI), vol. 10142, pp. 39–49. Springer, Cham (2017). https://doi.org/10.1007/978-3-319-51691-2_4
28. Krizhevsky, A.: Learning multiple layers of features from tiny images. Master's Thesis, Computer Science, University of Toronto (2009)
29. Krizhevsky, A., Sutskever, I., Hinton, G.: ImageNet classification with deep convolutional neural networks. In: Advances in Neural Information Processing Systems, pp. 1097–1105 (2012)
30. Barnsley, M.: Fractal Image Compression. AK Peters, Natick (1993)
31. Schmidhuber, J.: Low-complexity art. Leonardo **30**(2), 97–103 (1997)

Autonomy, Authenticity, Authorship and Intention in Computer Generated Art

Jon McCormack[(✉)] [iD], Toby Gifford[iD], and Patrick Hutchings[iD]

SensiLab, Faculty of Information Technology,
Monash University, Caulfield East, VC, Australia
Jon.McCormack@monash.edu
https://sensilab.monash.edu

Abstract. This paper examines five key questions surrounding computer generated art. Driven by the recent public auction of a work of "AI Art" we selectively summarise many decades of research and commentary around topics of autonomy, authenticity, authorship and intention in computer generated art, and use this research to answer contemporary questions often asked about art made by computers that concern these topics. We additionally reflect on whether current techniques in deep learning and Generative Adversarial Networks significantly change the answers provided by many decades of prior research.

Keywords: Autonomy · Authenticity · Computer art · Aesthetics · Authorship

1 Introduction: Belamy's Revenge

In October 2018, AI Art made headlines around the world when a "work of art created by an algorithm" was sold at auction by Christie's for US$432,500 – more than 40 times the value estimated before the auction [1]. The work, titled *Portrait of Edmond Belamy* was one of "a group of portraits of the fictional Belamy family"[1] created by the Paris-based collective *Obvious*.

The three members of Obvious had backgrounds in Machine Learning, Business and Economics. They had no established or serious history as artists. Their reasoning for producing the works was to create artworks "in a very accessible way (portraits framed that look like something you can find in a museum)" with the expectation of giving "a view of what is possible with these algorithms" [2].

The works' production involved the use of Generative Adversarial Networks (GANs), a technique developed by Ian Goodfellow and colleagues at the University of Montreal in 2014 [3]. It turned out that Obvious had largely relied on code written by a 19 year old open source developer, Robbie Barrat, who did not receive credit for the work, nor any remuneration from the sale (and who in turn, relied on code and ideas developed by AI researchers such as Goodfellow and

---

[1] The name is derived from the French interpretation of "Goodfellow": *Bel ami.*

© Springer Nature Switzerland AG 2019
A. Ekárt et al. (Eds.): EvoMUSART 2019, LNCS 11453, pp. 35–50, 2019.
https://doi.org/10.1007/978-3-030-16667-0_3

companies like Google). An online arts commentary site, Artnet, summarised it thus: "Obvious... was handsomely rewarded for an idea that was neither very original nor very interesting" [4].

During the auction, Barrat complained on twitter about the situation (Fig. 1):

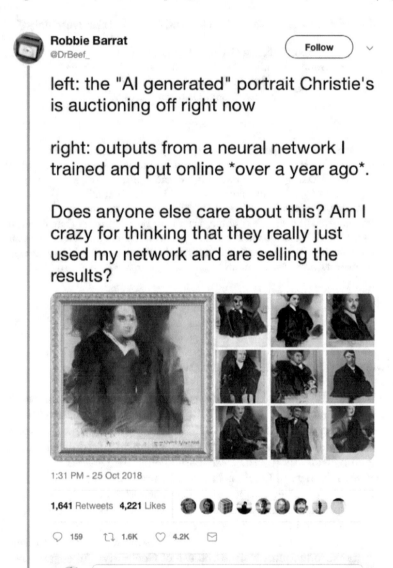

Fig. 1. Tweet by Robbie Barrat following the auction of *Portrait of Edmond Belamy*

Does anyone else care about this? Am I crazy for thinking that they really just used my network and are selling the results?

Following the sale of *Portrait of Edmond Belamy* to an undisclosed buyer, social media erupted with amazement, skepticism, and in some instances, outrage. Many publicly expressed concern that artists who had spent many years working with AI and art, and specifically GANs, had been overlooked; that their work represented far better examples of "AI Art" than Obvious' crude and derivative efforts. Ethical concerns were raised about Obvious being the "authors" of a work generated using code written by a third party (Barrat, whose code was based on Soumith Chintala's implementation of DCGANs [5]), even though Obvious were legally entitled to use it under the open source license under which it was released.

The sale of this work, and the attention it has subsequently brought to what was, until then, a relatively obscure area of research and creative production, raises many questions. Of course those with knowledge of history understand that creative applications of AI techniques date back to the earliest days of the use of the term "Artificial Intelligence" and even predate it.[2] The EvoMUSART workshop and conference series has been running since 2003 and has presented numerous examples of "art made by an algorithm".[3] The pioneering work of artists such as Michael Knoll, Georg Nees, Frieder Nake, Lilian Schwartz and Harold Cohen is widely acknowledged in both technical and artistic surveys of computer art and dates back to the early 1960s. Indeed, Cohen – a trained artist of some renown – spent most of his working life in the pursuit of AI Art, yet his works typically did not sell for even 1% of the price paid for *Portrait of Edmond Belamy*.

The sale and subsequent reaction to the work resurrects venerable questions regarding autonomy, authorship, authenticity, and intention in computer generated art. It should come as no surprise that, despite public reaction to the sale of *Portrait of Edmond Belamy*, these are well explored issues (see e.g. [7–20]). It is worth revisiting them for two reasons: firstly to encapsulate more than half a century of discussion around computer created art, and secondly, to see if recent advances in Machine Learning techniques, such as Deep Learning and GANs, change any of the well established research and commentary in this area.

1.1 Questions Raised

In this paper we review and comment on work in this area drawn from philosophy, psychology, cognitive science and computational creativity research, and use it to answer the following five key questions:

1. Who is the author of a work of art made by an AI algorithm?
2. What attribution or rights should be given to developers of an "art" generating algorithm?

---

[2] Ada Lovelace is famously known as one of the first people to record ideas about computer creativity [6].

[3] A great resource is the EvoMUSART Index, available at: http://evomusart-index.dei. uc.pt.

3. To what extent is any AI artwork relying on mimicry of human creative artefacts (rather than being independently creative)?
4. Is art made by "Deep Learning" techniques such as GANs different than art made by other generative algorithms?
5. Do artworks made by AI represent a new "kind" of art?

There are many additional interesting questions that could be considered in relation to AI art generally, and specifically in the case of *Portrait of Edmond Belamy*. For example, questions surrounding the extraordinary price paid for the work, the motivations of the creators, the quality of the work as an exemplar of the genre of AI Art, and so on. However answering these questions needs to draw on broader fields of research, such as market and social dynamics in the art world and auction system, human behavioural science and art criticism, and is beyond the scope of what is covered in this paper. A good introduction to these issues can be found in [21], for example.

1.2 AI Art

Throughout this paper we use the term "AI Art" in it's most obvious meaning: art that is made using *any* kind of AI technique. There is no real consensus on what is included or excluded from "AI Art". For example, evolutionary techniques are often considered a branch of AI, but traditionally, art made using evolutionary techniques is not referred to as AI Art. Many practitioners who adopt this term use deep learning techniques, such as GANs, Pix2Pix (Conditional Adversarial Networks), etc. For any cases in the paper where we need more specificity, we will refer to specific techniques.

2 Autonomy, Authenticity, Authorship and Intention (AAAI)

Art is a human enterprise, defined by experience, of both artist and audience. Demarcations of the continuous actions and/or material manipulations of an artist into art-objects (be they paintings, sculptures, poems, scripts, scores or performances) take place in myriad ways, though commonly through the intentional declaration of an artist that a particular artefact is to be considered a finished work. Whether the result of years of meticulous crafting, such as Michelangelo's *David*, or readymades such as Duchamp's *Fountain*, the "trait inherent in the work of the artist [is] the necessity of sincerity; the necessity that he shall not fake or compromise" [22]. The artist's intention, then, is critical, and intertwined with issues a work's authenticity. Even more fundamental is the autonomy of the artist and their authorship when considering the aesthetic value of an artefact. Thus when considering work generated, in part or whole, by an artificial intelligence, the extent to which these properties can be ascribed to the machine is brought into question.

2.1   Autonomy

One of the main attractions of working with generative computational systems is their capacity for agency and autonomy. Systems that can surprise and delight their authors in what they produce are undoubtedly an important motivation for working with computer generated creative systems [23].

Boden provides an extensive examination of autonomy in computer art [17, Chapter 9]. She distinguishes between two different kinds of autonomy in non-technological contexts and their parallels in computational art:

– physical autonomy such as that exhibited in homeostatic biological systems, and
– mental/intentional autonomy typified by human free-will.

Boden describes how these different kinds of autonomy are at play in generative art. Many concepts used in generative art, such as emergence, agency, autopoesis, homoeostasis and adaptation are underpinned by *self-organisation*; something Boden views as synonymous with a specific kind of autonomy where, "the system's independence is especially strong: it is not merely self-controlled, but also self-generating" [17, p. 180], with the "self" in self-organisation referring to the impersonal components of the system, not the intentional, mental self.

Boden's second form of autonomy is inherently tied to human freedom: something lacking in the autonomy of AI art systems that use an emergent, "bottom up" philosophy (such as Artificial Life or Evolutionary based systems). This intentional autonomy requires concepts from the Theory of Mind, such as intention, desire and belief. There is nothing in principle that dismisses the possibility of artificial creative systems possessing this intentional autonomy, and indeed, this is something extensively explored by Artificial Intelligence research since its inception. If this class of autonomy is dependent on free-will, then any artificially autonomous creative software would potentially need to make the choice to make art, not be explicitly programmed to do so.

Autonomy of Deep Learning Systems. With this general distinction in mind, it is worth considering how deep learning techniques, such as GANs fit into these different categories of autonomy. A GAN was used by Obvious to create *Portrait of Edmond Belamy*. GANs use two competing neural networks, a *generator* and a *discriminator*. The generator learns to map from a latent space to some (supplied) data distribution, while the discriminator attempts to discriminate between the distribution produced by the generator and the real distribution.[4]

Perceptions of autonomy in GANs rest on their ability to generate convincing "fakes" (convincing to the discriminator at least). They have autonomy in the sense of being able to synthesise datasets that mimic the latent space of their

---

[4] This is conceptually similar to co-evolutionary strategies, such as creator/critic systems [24], used at least since the 1990s.

training data – in many current artistic works this training data comes from digitised works of other artists, often those recognised in the cannon of classical or modernist art. Certainly they have a very limited autonomy in choosing what statistical model might best represent the latent space and in the datasets generated. In visual examples, this is often akin to structured "mashups": images with similar overall appearances to parts of images from the training set, recalling the visually similar, but algorithmically different technique of *morphing* [25], developed in the early 1990s.

In contrast, human artists do not learn to create art exclusively from prior examples. They can be inspired by experience of nature, sounds, relationships, discussions and feelings that a GAN is never exposed to and cannot cognitively process as humans do. The training corpus is typically highly curated and minuscule in comparison to human experience. This distinction suggests GANism is more a process of mimicry than intelligence. As flowers can utilise Dodsonian mimicry to lure pollinators by generating similar sensory signals [26], current GAN techniques lure human interest by mimicking valued aesthetic properties of prior artistic artefacts without being a direct copy [27–30].

At a technical level, GANs are a technique for finding a statistical distribution of latent dimensions of a training set. Hence their capacity for autonomy is limited, certainly no more that prior machine learning or generative systems. Even in their perceived autonomy as image creators, their ability to act autonomously is limited within a very tight statistical framework that is derived from their training data. There is no evidence to suggest that they posses any of Boden's second sense of intentional autonomy.

Thus any claims we can make about the autonomy of a GAN-based software system's autonomy are limited. Certainly many different generative systems with equal or greater autonomy exist (in the non-intentional sense). While a claim such as, "an AI created this artwork" might be literally true, there is little more autonomy or agency that can be attributed to such an act than would be to a situation where "a word processor created a letter", for example.

Summary. Artistic generative systems illuminate the complex relationships between different kinds of autonomy. Artists and researchers have worked with such systems because they value their ability to exploit autonomous (in the self-organising, homeostatic sense) processes, while often (incorrectly) ascribing autonomy in the intentional sense to them. Nonetheless, autonomous generative processes result in more aesthetically challenging and interesting artworks over what can be achieved when the computer is used as a mere tool or "slave". They open the possibility for types of expression not possible using other human tools. However, that does not give them the intentional autonomy characterised by human creativity.

2.2   Authorship

Issues of agency and autonomy also raise another important concept challenged by generative art: authorship. Who is the author of an artwork that is generated by a computer program? A computer program that can change itself in response to external stimuli has the potential to learn and adapt (autonomy in the self-organising sense). Hence it can do things that the programmer never anticipated or explicitly designed into the software, including the potential to act creatively, even if this ability was not part of its original programming (see e.g. [31]). Thus while we may describe the programmer as the author of the program, it seems that our commonsense notion of authorship should extend to recognising the program itself as at least a partial author of the artwork in such circumstances. Commonsense, however, may be culturally determined and historically fraught; and the notion of authorship has a vexed and contested history.

Authorship and Individuality. To talk of an author presupposes the notion of an *individual*; a notion that only developed into something close to its current form in western philosophy in the 17th century, and even today has different forms in many eastern cultures. Take Thomas Hobbes' 1651 definition of a person: "He whose words or actions are considered either as his own or as representing the words or actions of any other thing to whom they are attributed, whether truly or by fiction" [32]. Hobbes distinguishes one whose words are considered their own as a *natural person* and one whose words are considered as representing another an *artificial person*. Only natural persons does Hobbes deign 'authors'.

Hobbes' definition of a person appears to preclude female persons within the very first word. This highlights the relativism of notions such as person-hood and authority to the ambient culture. The question of whether or not a computer could ever be considered an "author" may be tractable only to cultural argument, rather than relating to either objective truth or universal human experience.

An instructive case study is that of Albert Namatjira, one of the first Indigenous Australian artists to adopt a European landscape style in his paintings. So enamoured were the colonial powers of this "civilised" Aborigine (who even met the Queen) that Albert was honoured as the first Indigenous Australian to be granted citizenship of Australia in 1957. Here we see perhaps the person-author relationship operating inversely to the received wisdom: his artistic authorship (in the culturally approved style) propelled him from non-entity into person-hood.

When Namatjira died he bequeathed his copyright (possession of which was one perk of being considered a natural person) to his family, and subsequently his paintings have become enormously valuable. Unfortunately his family did not enjoy person-hood, and so the bequest was managed by the state, which chose (without consultation) to sell all rights to a private investor for a pittance. The family has been fighting ever since for the copyright to be returned to their community, and have only recently succeeded in re-obtaining it.

This story is by no means unique, nor restricted to the visual arts. The music world is replete with examples of traditional musics being co-opted by colonisers, recorded and copyrighted, the original performers subsequently persecuted for continuing to perform it. The notion of authorship here is simply an expression of power. As times progress, with womens' suffrage, Indigenous citizenship, and increasing egalitarianism so too the granting of authorship has expanded. Perhaps when AIs are granted citizenship they too will receive legal recognition as authors, although the first such citizen robot – Sophia – has yet to produce any creative works. More likely, as with Albert Namatjira, the causality will flow the other way; when AIs start autonomously creating good art (in the culturally accepted style), then robots will be granted status as people.

Authorship and Romanticism. Our current understanding of authorship is rooted in the Romantic movement. Whilst contemporary culture has abandoned many Romantic ideas, including that of the lone genius, the concept of authorship has remained relatively fixed, probably due to the development of copyright and intellectual property laws in the late Enlightenment and early Romantic period, and the ensuing substantial vested interests in its perpetuation. Jaszi [33] suggests "it is not coincidental that precisely this period saw the articulation of many doctrinal structures that dominate copyright today. In fact British and American copyright present myriad reflections of the Romantic conception of 'authorship'". For example, in a mid 20th century legal case an artist superimposed two images from the *Wizard of Oz* and attempted to claim copyright on the result (preempting copyright disputes that would follow over mashups and remixes). The judge rejected that application saying "we do not consider a picture created by superimposing one copyrighted photographic image on another to be original" [33]. Jaszi argues this decision has no real basis in the legal arguments made, but that:

> this decision does make sense, however, when viewed in light of the Romantic "authorship" construct, with its implicit recognition of a hierarchy of artistic productions. In that hierarchy, art contains greater value if it results from true imagination rather than *mere application*, particularly if its creator draws inspiration directly from nature [33].

Echoes of this debate continue in the computational creativity field, surfacing in the contest between critics of *mere generation* [34] and computational media artists for whom computer generated artefacts (be they complete, partial, or co-constructed) form an integral part of their practice [35].

Legal and Moral Ownership. The concept of authorship, as affected by non-anthropocentric intelligence and cognitive labour is also of interest in legal discussions around intellectual property, and has been for several decades. In 1997 Andrew Wu [36] wrote on issues of copyright surrounding computer-generated works. He perceived a gap between the objective of copyright and the idea of copyrighting works produced by an intelligent computer.

> ... the basic problem presented by intelligent computers; awarding copyright to the one who is the author–in the sense of being the originator or intellectual inventor of a work–does not further the objective of stimulating future creativity [36].

Copyright is intended to encourage discovery and creation by reducing the fear that creators might have of others benefiting from their work instead of the creators themselves. Currently, computers don't have the economic needs, emotional concerns or desires that drive copyright legislation, and so the incentive to provide copyright to software systems is greatly reduced. At the same time there is an awareness that software can have a significant role in the creation of artefacts that are deemed to be creative works.

While legal interpretations will differ between countries, it is clear that artificial intelligence presents unique challenges for the concept of copyright and was a point of discussion for legal scholars in the last century.

In 1984 the US Copyright Office listed a computer program named "Racter" as an author of *The Policeman's Beard is Half Constructed* but not a claimant of the copyright [37]. The distinction between claimant and author is interesting because it contrasts the modern construct of companies as economically driven legal entities that can claim ownership over physical and virtual goods, but are not deemed to be authors of any artefact themselves.

Today the same questions are still being raised around authorship for copyright. In arguing for patent protection for computer generated works (CGWs), Ryan Abbott suggests that legal changes to copyright may have been slowed by a lack of commercial need:

> Given these technological advances, one would be forgiven for asking— where are the CGWs? Why are there not routinely lawsuits over CGWs? How have countries managed without legal standards for CGWs? It may be that the creative AI revolution has yet to arrive. CGWs may be few and far between, or lack commercial value [38].

Now that artworks made with generative systems have resulted in significant commercial gains there may be increased interest in establishing legal standards around them. There may also be a shift in the dynamics of AI art as a movement and community. Until now, artists have both relied on and contributed to the distribution of open-source code. The benefits of community and reputation building by sharing open-source code have outweighed potential economic returns for AI art, but have left little basis for claims of legal ownership of these systems.

Without legal ownership, authors maintain a moral right to partial ownership of the process. This ownership can be expressed in the naming of a technique or algorithm, or honest and accurate attribution. Software and datasets, especially those utilising creative works of other individuals, are attributable contributions to machine learning-based systems.

Summary. Ascribing authorship of a creative work is only meaningful in a particular cultural and legal context. In a contemporary Western context, authorship of AI generated art will be shared between the artist, the developer of the underlying algorithms, and possibly any number of contributors to the training data whose style is being abstracted. Increasingly the algorithms themselves may be accepted as partial authors, both morally and legally, in step with their increasing perception as autonomous intelligent agents. However, the level of intelligence of existing AI systems falls short of its perception in current popular imagination. As such, greater attention should be focused on accurate and inclusive attribution amongst the constellation of human actors. Accurate attribution not only benefits these authors, but helps establish the authenticity of work produced with AI systems.

2.3   Authenticity

The authenticity of computer art often provokes quite binary responses. A number of theorists, critics and philosophers rule out the idea of computer art entirely being authentic because the art was made by a machine, not a person [14].

A common criticism of works employing algorithms is that they are *generic*, and such genericism implies that they lack authenticity. As was the case with *Portrait of Edmond Belamy*, the human creators relied on software derived from a common pool of algorithms and code, and the Obvious team played only a very minor role in actually even writing any software [39]. The majority of algorithms used for computer art have originated from the sciences, not the arts, so understandably they evoke scepticism when introduced as key components in an artistic process [40].

These and similar criticisms can be characterised by three related points:

1. That works made with the same or similar algorithms – even by different artists – possess a certain generic and repetitive character;
2. Artists do not fully understand, or are misleading about, the process of creation, in that the algorithms do not exhibit the characteristics or outcomes that the artists' claim the processes used represent;
3. Works exclude artistic possibilities due to their dependence on a generic technical process that arose in an unrelated context.

Addressing the first point, it is typically the case that similar algorithms produce similar results. Repeatability and determinism are fundamental utilities of computation after all. And if the algorithm is actually the "artist" then it makes sense that the same algorithm would produce works that are the same or with highly similar characteristics.[5] There is nothing inherently so generic in creative software in general – a tool like Photoshop is theoretically capable of creating any style or kind of image. This is because most of the *creative agency* rests with the user of the software, not the software itself [16]. Machine learning researcher, Francois Chollet has referred to art made by GANs as *GANism*[6], implying the algorithm is responsible for the style more than the artists who use the software. As deep learning software becomes increasingly complex and difficult to understand, authorship and creative agency shifts towards the algorithm itself. However, in the case of reinforcement learning, training data plays

---

[5] It is worth noting that artistic styles or idioms share similar characteristics too.
[6] https://twitter.com/fchollet/status/885378870848901120.

a crucial role in determining what the learning system does. When the training data is authored by others, they implicitly make a contribution, akin to the way that mashups, collage or sampling have contributed to artworks previously (see Sect. 2.2).

Algorithmically generated art systems emphasise process as the primary mechanism of artistic responsibility. A generic or copied process will produce generic work. In considering generative systems in an artistic context, the "Art" is in the construction of process. If that process is derivative or memetic then the work itself will likely share similar properties to others, making any claims to artistic integrity minimal.

Truth About Process. Outside of technical research circles, terms such as "Artificial Intelligence" inherently carry assumed meaning, and the value of this fact has not escaped numerous entrepreneurs and marketing departments who claim their work has been "created by an AI." People will naively find anything associated with non-human "intelligence" interesting, simply because we aren't used to seeing it in the wild. But because the algorithms and techniques are largely opaque to non-experts, it is natural – but incorrect – to assume that an artificial intelligence mirrors a human intelligence (or at least some major aspects of human intelligence). Such views are reinforced by the anthropomorphic interfaces often used for human-AI interaction, such as realistic human voices, scripted personalities, etc.

Recent analysis of public perception of AI indicates that non-experts are more generally optimistic about the possibilities of AI than experts [41]. In one survey 40% of respondents "think AI is close to fully developed" [42]. Such surveys reinforce the problem of dissonance between public perceptions and reality regarding technological capability.

> We don't ascribe artistic integrity to someone who produces art in indefinitely many styles on request [17, p. 187].

Any full disclosure of artistic process in AI art needs to appreciate the authenticity of the generative process as fundamental to the artwork. Current AI Art, including that made using GANs, are *not* intelligent or creative in the way that human artists are intelligent and creative. In our view, artists and researchers who work with such techniques have a responsibility to acknowledge this when their work is presented and to correct any misunderstandings where possible.

2.4   Intention

Another well discussed issue in relation to computer generated art is that of *intention*. How can a software program be considered an "artist" if it lacks the independent intention to make art or to *be* an artist? Software artists are employed like slaves: born into servitude of their creators to relentlessly and unquestionably be art production machines. We would never force a human intelligence to only be concerned with producing art, nor would we consider an artist just an art-production machine, yet we have no difficulty saying that artificial intelligence must serve this role.

As discussed in Sect. 2.1, the kind of autonomy typified by free-will that human artists possess includes the ability to decide not to make art or to be an artist. Software programmed to generate art has no choice but to generate; it lacks "the intentional stance" [43] of mental properties and states that might lead to a consideration of making art in the first place. So in addition to lacking autonomy, AI Art also lacks intention. Such conditions are often glossed over in everyday descriptions, where anthropomorphism and personalising are commonplace.

The terminology surrounding algorithmically generated art systems has shifted in recent years. While deep neural networks share many mathematical properties with Markov models and can employ evolutionary algorithm techniques for training, systems utilising deep learning have quickly gained the label of "artificial intelligence" while existing techniques are typically labelled "generative". Intelligence has a much stronger association with intent than does generation, and the shift in terminology is likely affecting the perception of intention in these systems.

3   Questions Answered

Having now examined the issues of autonomy, authenticity, authorship and intention, we are now able to answer the questions posed in Sect. 1.1.

3.1   Who Is the Author of a Work of Art Made by an AI Algorithm?

The creator of the software and person who trained and modified parameters to produce the work can both be considered authors. In addition to these contributors, artists whose works feature prominently in training data can also be considered authors. There has been precedent cases of software systems being considered legal authors, but AI systems are not broadly accepted as authors by artistic or general public communities. See Sect. 2.2.

3.2   What Attribution or Rights Should Be Given to Developers of an "Art" Generating Algorithm?

Authors have a responsibility to accurately represent the process used to generate a work, including the labour of both machines and other people. The use of open-source software can negate or reduce legal responsibilities of disclosure, but the moral right to integrity of all authors should be maintained. See Sect. 2.3.

3.3   To What Extent Is Any AI Artwork Relying on Mimicry of Human Creative Artefacts?

AI systems that are trained to extract features from curated data-sets constructed of contents produced by people are relying on mimicry of artefacts rather than autonomously searching for novel means of expression. This holds true for current, popular AI art systems using machine learning. See Sect. 2.1.

3.4 Is Art Made by "Deep Learning" Techniques Such as GANs Different Than Art Made by Other Algorithms?

There are no significant new aspects introduced in the process or artefact of many GAN produced artworks compared to other established machine learning systems for art generation. Currently, there is a difference in the way GANs are presented by media, auction houses and system designers: as *artificially intelligent systems* that is likely affecting the perception of GAN art. As this difference is grounded more in terminology and marketing than intrinsic properties of the technique, history suggests it is not likely to sustain. In her 1983 paper clarifying the terminology of "Generative Systems" and "Copy Art" Sonia Sheridan stated:

> Despite our efforts, "Copy Art" emerged to exploit a single system for a marketable, recognizable art product. In our commercial society and in the context of conditions prevailing through the seventies, such a development was inevitable, as was the struggle to maintain the character and integrity of the Generative Systems approach [44].

See Sects. 1.2 and 2.4.

3.5 Do Current Artworks Made by AI Represent a New "Kind" of Art?

Probably not in any major way. At least no more than any other kind of computer generated art. And let us not forget that computer generated art is now more than 50 years old. Certainly there are some stylistic differences between GAN Art and other types of computer generated art, just as there are many different painting styles. However, the idea of a computer autonomously synthesising images isn't new. Nor are the conceptual ideas or process presented in contemporary AI Art new. It might be argued that the technical methodology is new, even if it is largely developed by non-artists for non-artistic purposes. However, differences in technical methods rarely constitute a new kind of art on their own – cultural and social factors typically play a far greater role.

Summary. We hope this paper has helped clarify and reiterate some of the large body of research around the status of computer generated art, particularly those works that claim to be authored, in full or in part, by non-human intelligence. We leave it to the reader to make judgement on the merits or otherwise of the sale of *Portrait of Edmond Belamy* and the claims made by Obvious and Christies.

Acknowledgements. This research was support by Australian Research Council grants DP160100166 and FT170100033.

References

1. Is artificial intelligence set to become art's next medium? November 2018. https://www.christies.com/features/A-collaboration-between-two-artists-one-human-one-a-machine-9332-1.aspx. Accessed 07 Nov 2018

2. Obvious: Obvious, explained, February 2018. https://medium.com/@hello.obvious/ai-the-rise-of-a-new-art-movement-f6efe0a51f2e. Accessed 01 Nov 2018

3. Goodfellow, I., et al.: Generative adversarial nets. In: Advances in Neural Information Processing Systems, pp. 2672–2680 (2014)

4. Rea, N.: Has Artificial Intelligence Brought Us the Next Great Art Movement? Here Are 9 Pioneering Artists Who Are Exploring AI's Creative Potential, November 2018. https://news.artnet.com/market/9-artists-artificial-intelligence-1384207. Accessed 07 Nov 2018

5. Radford, A., Metz, L., Chintala, S.: Unsupervised representation learning with deep convolutional generative adversarial networks. Preprint, arXiv:1511.06434 (2015)

6. Menabrea, L.F.: Sketch of the analytical engine invented by Charles Babbage, No. 82, Bibliothèque Universelle de Genève, October 1842. http://www.fourmilab.ch/babbage/sketch.html

7. Burnham, J.: Systems esthetics. Artforum **7**, 30–35 (1968)

8. Burnham, J.: On the future of art. In: Fry, E.F. (ed.) The Aesthetics of Intelligent Systems, p. 119. The Viking Press, New York (1969)

9. Reichardt, J.: The Computer in Art. Studio Vista; Van Nostrand Reinhold, London, New York (1971)

10. Davis, D.: Art and the Future. Praeger, New York (1973)

11. Leavitt, R.: Artist and Computer. Harmony Books, New York (1976)

12. Dietrich, F., Tanner, P.: Artists interfacing with technology: basic concepts of digital creation, July 1983

13. McCorduck, P.: Aaron's Code: Meta-art, Artificial Intelligence and the Work of Harold Cohen. W.H. Freeman, New York (1990)

14. O'Hear, A.: Art and technology: an old tension. R. Inst. Philos. Suppl. **38**, 143–158 (1995)

15. Cohen, H.: The further exploits of Aaron, painter. Stanford Humanit. Rev. **4**, 141–158 (1995). http://www.stanford.edu/group/SHR/4-2/text/cohen.html

16. Bown, O., McCormack, J.: Creative agency: a clearer goal for artificial life in the arts. In: Kampis, G., Karsai, I., Szathmáry, E. (eds.) ECAL 2009. LNCS (LNAI), vol. 5778, pp. 254–261. Springer, Heidelberg (2011). https://doi.org/10.1007/978-3-642-21314-4_32

17. Boden, M.A.: Creativity and Art: Three Roads to Surprise. Oxford University Press, Oxford (2010)

18. McCormack, J.: Aesthetics, art, evolution. In: Machado, P., McDermott, J., Carballal, A. (eds.) EvoMUSART 2013. LNCS, vol. 7834, pp. 1–12. Springer, Heidelberg (2013). https://doi.org/10.1007/978-3-642-36955-1_1

19. McCormack, J., Bown, O., Dorin, A., McCabe, J., Monro, G., Whitelaw, M.: Ten questions concerning generative computer art. Leonardo **47**(2), 135–141 (2014)

20. d'Inverno, M., McCormack, J.: Heroic vs collaborative AI for the arts. In: Yang, Q., Wooldridge, M. (eds.) Proceedings of the Twenty-Fourth International Joint Conference on Artificial Intelligence (IJCAI 2015), pp. 2438–2444. AAAI Press (2015)

21. Thornton, S.: Seven Days in the Art World. Granta, London (2009)

22. Dewey, J.: Art as Experience. Capricorn Books, New York (1934)

23. McCormack, J.: Working with generative systems: an artistic perspective. In: Bowen, J., Lambert, N., Diprose, G. (eds.) Electronic Visualisation and the Arts (EVA 2017), pp. 213–218. Electronic Workshops in Computing (eWiC), BCS Learning and Development Ltd., London, July 2017

24. Todd, P.M., Werner, G.M.: Frankensteinian methods for evolutionary music composition. In: Griffith, N., Todd, P.M. (eds.) Musical Networks: Parallel Distributed Perception and Performance, pp. 313–339. The MIT Press/Bradford Books, Cambridge (1999)

25. Beier, T., Neely, S.: Feature-based image metamorphosis. Comput. Graph. **26**, 35–42 (1992)

26. Dodson, C.H., Frymire, G.P.: Natural Pollination of Orchids. Missouri Botanical Garden, St. Louis (1961)

27. Tan, W.R., Chan, C.S., Aguirre, H.E., Tanaka, K.: ArtGAN: artwork synthesis with conditional categorical GANs. In: 2017 IEEE International Conference on Image Processing (ICIP), pp. 3760–3764. IEEE (2017)

28. He, B., Gao, F., Ma, D., Shi, B., Duan, L.Y.: ChipGAN: a generative adversarial network for Chinese ink wash painting style transfer. In: 2018 ACM Multimedia Conference on Multimedia Conference, pp. 1172–1180. ACM (2018)

29. Zhang, H., Dana, K.: Multi-style generative network for real-time transfer. Preprint, arXiv:1703.06953 (2017)

30. Elgammal, A., Liu, B., Elhoseiny, M., Mazzone, M.: CAN: creative adversarial networks, generating "art" by learning about styles and deviating from style norms. Preprint, arXiv:1706.07068 (2017)

31. Keane, A.J., Brown, S.M.: The design of a satellite boom with enhanced vibration performance using genetic algorithm techniques. In: Parmee, I.C. (ed.) Conference on Adaptive Computing in Engineering Design and Control 96, pp. 107–113. P.E.D.C. (1996)

32. Hobbes, T.: Leviathan. In: Molesworth, W. (ed.) Collected English Works of Thomas Hobbes. Routledge, London (1651/1997)

33. Jaszi, P.: Toward a theory of copyright: the metamorphoses of authorship. Duke Law J. **40**(2), 455 (1991)

34. Veale, T.: Scoffing at mere generation (2015). http://prosecco-network.eu/blog/scoffing-mere-generation. Accessed 18 Aug 2015

35. Eigenfeldt, A., Bown, O., Brown, A.R., Gifford, T.: Flexible generation of musical form: beyond mere generation. In: Proceedings of the Seventh International Conference on Computational Creativity, pp. 264–271 (2016)

36. Wu, A.J.: From video games to artificial intelligence: assigning copyright ownership to works generated by increasingly sophisticated computer programs. AIPLA QJ **25**, 131 (1997)

37. Chamberlain, W.: Copyright catalog: the policeman's beard is half constructed: computer prose and poetry. https://cocatalog.loc.gov/cgi-bin/Pwebrecon.cgi?SC=Author&SA=Racter&PID=zQyq43DuiIh6HvQpRNEOeM8BmyRRN&BROWSE=1&HC=1&SID=2. Accessed 08 Nov 2018

38. Abbott, R.: Artificial Intelligence, Big Data and Intellectual Property: Protecting Computer-Generated Works in the United Kingdom. Edward Elgar Publishing Ltd., Cheltenham (2017)

39. Vincent, J.: How three French students used borrowed code to put the first AI portrait in Christie's (2018). https://www.theverge.com/2018/10/23/18013190/ai-art-portrait-auction-christies-belamy-obvious-robbie-barrat-gans. Accessed 08 Nov 2018

40. Parikka, J.: Leonardo book review: the art of artificial evolution: a handbook on evolutionary art and music. http://www.leonardo.info/reviews/nov2008/parikka_art.html. Accessed 08 Feb 2019

41. Manikonda, L., Kambhampati, S.: Tweeting AI: perceptions of lay versus expert twitterati. In: Proceedings of the Twelfth International AAAI Conference on Web and Social Media (ICWSM 2018), pp. 652–655 (2018)

42. Gaines-Ross, L.: What do people - not techies, not companies - think about artificial intelligence? Harvard Bus. Rev. (2016). https://hbr.org/2016/10/what-do-people-not-techies-not-companies-think-about-artificial-intelligence

43. Dennett, D.C.: The Intentional Stance. MIT Press, Cambridge (1987)

44. Sheridan, S.L.: Generative systems versus copy art: a clarification of terms and ideas. Leonardo **16**(2), 103–108 (1983)

Camera Obscurer: Generative Art for Design Inspiration

Dilpreet Singh[1], Nina Rajcic[1], Simon Colton[1,2(✉)], and Jon McCormack[1]

[1] SensiLab, Faculty of IT, Monash University, Melbourne, Australia
simon.colton@monash.edu
[2] Game AI Group, EECS, Queen Mary University of London, London, UK

Abstract. We investigate using generated decorative art as a source of inspiration for design tasks. Using a visual similarity search for image retrieval, the *Camera Obscurer* app enables rapid searching of tens of thousands of generated abstract images of various types. The seed for a visual similarity search is a given image, and the retrieved generated images share some visual similarity with the seed. Implemented in a hand-held device, the app empowers users to use photos of their surroundings to search through the archive of generated images and other image archives. Being abstract in nature, the retrieved images supplement the seed image rather than replace it, providing different visual stimuli including shapes, colours, textures and juxtapositions, in addition to affording their own interpretations. This approach can therefore be used to provide inspiration for a design task, with the abstract images suggesting new ideas that might give direction to a graphic design project. We describe a crowdsourcing experiment with the app to estimate user confidence in retrieved images, and we describe a pilot study where Camera Obscurer provided inspiration for a design task. These experiments have enabled us to describe future improvements, and to begin to understand sources of visual inspiration for design tasks.

1 Introduction and Motivation

Producing decorative abstract images automatically can be achieved in numerous ways, e.g., via evolutionary computing [1] or generative adversarial networks [2]. With modern computing power, it's possible to produce tens of thousands of images of fairly high resolution in a short time, and driven by human interaction and/or a suitable fitness function, these images could have high aesthetic appeal and diversity. Such generated images can be incorporated into artistic practice through careful curation and exhibition [3], and such images have found other uses, for example as the basis for scene generation [4]. However, it is fair to say that there are diminishing returns when it comes to the value of large databases of generated abstract decorative images. One of the motivations of the work presented here is to find an interesting usage for large databases of such generated images, namely to provide inspiration for design projects.

ⓒ Springer Nature Switzerland AG 2019
A. Ekárt et al. (Eds.): EvoMUSART 2019, LNCS 11453, pp. 51–68, 2019.
https://doi.org/10.1007/978-3-030-16667-0_4

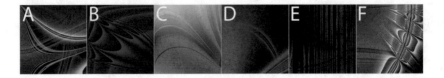

Fig. 1. Interpretable decorative images generated by $(x, y) \to (r, g, b)$ functions.

Abstract art pieces made by people often offer various opportunities for interpretation and/or self reflection. In the eye of a particular viewer, an abstract image could bear a resemblance to an object, person, animal, texture or place; it could bring back good or bad memories; evoke an emotion; set a mood or reflect certain aesthetic considerations like symmetry, balance, depth and composition. Generated abstract images similarly offer possibilities for interpretation. As people are very good at interpreting meaning in images constructed with none in mind, generated images can often reflect objects, showcase aesthetics, evoke emotions and possibly provide inspiration for tasks.

As an example, the images in Fig. 1 were constructed using a very straightforward approach dating back to Sims [5]. Here, algorithms which take (x, y) coordinates as input and output triples of (r, g, b) values using trigonometric, arithmetic and exponential functions were automatically generated and run to produce the images. Even though the process for generating these images is simplistic (in generative art terms), the images can still be evocative. For instance, image A may remind a viewer of a complicated motorway junction, while image B may evoke memories of Halloween ghosts. Image C has a look of light reflecting off a shiny surface, while image D looks a little like a feather quill being used on paper. Of images E and F, most people would agree that one could be interpreted as evoking a sinister mood, while the other evokes a fiesta mood.

Some projects have aimed to produce images which invite interpretation in certain ways, e.g., Todd and Latham's Mutator project originally aimed to produce images resembling organic forms [6], and McCormack produced artworks depicting imaginary flower and plants evocative of Australian flora [7]. Machado et al. have evolved abstract pieces for various interpretations including as faces [8] and figures [9]. An alternative to generating images for a particular interpretative purpose is to generate large databases of abstract images and use the fact that many will have reasonable visual interpretations (albeit somewhat subjectively). The images in Fig. 1 were chosen by hand specifically because of places, objects or moods they might evoke in an average viewer. Automating this skill of visually identifying images with valid interpretations could be useful, as it would enable mass-generation of abstract decorative images in an unfocused way, with a user providing focus later on, e.g., looking for images evoking ghosts when the deadline for a Halloween graphic design project is looming.

An important question when implementing such an approach is how to retrieve abstract images from a large database. There are many potential solutions to this, and we investigate here using visually similar search, where the user supplies a *seed image* for image retrieval, and the app tries to find visually matching images from various datasets. The seed image could come from a variety of sources such as a user's surroundings, a hand-drawn sketch or an image on their computer screen. To test this approach, we have developed a mobile application called *Camera Obscurer* which retrieves images from two large datasets of generated images, given a seed image for a visual similarity search. As described in Sect. 2, the two datasets are: (a) around 9,500 abstract art images similar in nature to those in Fig. 1 but produced by a different method described below, and (b) 20,000 filtered images, obtained by applying 1,000 generated image filters to 20 digital photos representative of the real world.

Running on a camera-enabled mobile phone or tablet, the system allows the user to take a photograph from their surroundings with their device, instantly receive images from the datasets and upload some to a Pinterest board as they see fit. The first expected usage for Camera Obscurer is simply to find images valuable to a graphic design project, i.e., that might be used in the project. Hence, if a user had a ghost theme in mind for a Halloween design, and wanted an abstract artwork somewhat evocative of ghosts, they could (a) photograph something nearby which looks a little like a ghost (b) draw something ghost-like or (c) bring up an image in their browser of a ghost and photograph that. In principle, if a ghost-like image is available in the database, it should be retrieved. However, while the effectiveness of visual similarity search has been evaluated with respect to retrieving similar digital photographs of real-world images, abstract images don't in general hugely resemble real-world seed images. Hence, to test whether abstract images can be sensibly retrieved, we performed a crowdsource experiment to measure whether retrieved images resemble the seed images more so than a randomly chosen one, as described in Sect. 3.1.

A welcome side-effect of abstract images having relatively low visual similarities with seed images is that the retrieved images may have numerous visual interpretations, e.g., as objects different from those in the seed image, or with different textures, juxtapositions and/or evoking different moods and emotions. Hence a retrieved abstract image could supplement a user's seed image and provide inspiration. The second expected usage for Camera Obscurer is to provide higher-level inspiration for design projects, over and above the retrieval of potentially useful images. Imagine, for instance, someone undertaking a design project and wanting to employ a texture, but unsure of which to use. Here, they could use the app to photograph textures around them to use as seeds, and inspect any retrieved images resembling textures and hopefully make a more informed choice. This approach may be useful for amateurs undertaking graphic design projects with little design background and/or starting with little or no inspiration.

To investigate the potential of this approach for design inspiration, we undertook a pilot study where 8 participants used Camera Obscurer to help with a design project. To compare and contrast the value of abstract generated pieces, the app also retrieved colour palettes, images from a photo archive and images of human-made art. As described in Sect. 3.2, the purpose of the pilot study was to raise suitable questions about using abstract retrieved images for design inspiration, rather than to test particular hypotheses. We conclude with these questions in Sect. 4, and suggest improvements for design tasks by: enabling generative artworks and filters to be changed as part of a design workflow; and enabling the software to provide textual interpretations of abstract pieces.

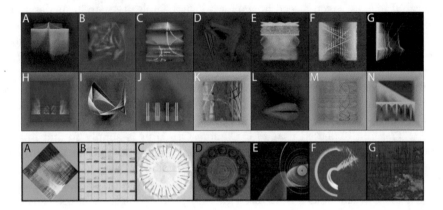

Fig. 2. (i) Abstract generated images with interpretable depth (B, H, J, L), patterns (E, F, K), motion (C, G, M), lighting effects (C, H, N), textures (B, E, F, M), objects (D, I, L) and locations (A, C, J, N). (ii) Images produced with (A) diamond (B) grid (C) kaleidoscope (D) necklace (E) oval (F) polar and (G) tron transformations.

2   The Camera Obscurer System

2.1   Image Curation

To recap, our intention is to test whether abstract generated imagery retrieved in response to a seed image from a user's surroundings could provide visual inspiration. This needs a large dataset of decorative art images which (a) is varied enough to include images having some visual similarity with a large range of seeds and (b) contains images which can be interpreted in various ways. To this end, we used an archive of generated decorative images previously described in [10] and a dataset of generated image filters, as described in [11].

The decorative images were produced by evolving a series of 10 functions which take as input a particle number, p, between 1 and 1,000, a timestep number, t, between 1 and 100, and previous values of any of the functions. Each function outputs numerical values which are interpreted as (a) initialisation functions $i_1, \ldots, i_5$ calculating (x, y) coordinates and (r, g, b) colours of 1,000 particles and (b) updating functions $u_1, \ldots, u_5$ calculating the coordinates and colour of a particle at timestep t. As each function can use previously calculated values, the calculations performed become highly iterative and complex. To use the functions, an image of a given size is first generated as a monotone background, with the colour chosen randomly. Then 1,000 particles are initialised via i_1 to i_5 and then moved and changed (in colour) with u_1 to u_5 over 100 timesteps. At each timestep, a line is drawn between each particle's previous position and its new one, with the line rendered in the new colour of the particle. After each timestep, a blur is also applied to the entire image.

Fig. 3. Image filter examples: original image on the left, 7 filtered versions on the right.

We have found the images generated in this fashion to contain highly varied, aesthetically pleasing (albeit subjectively) decorative examples, and we have performed a number of experiments to search the space, using fitness functions [4] and human guidance [10]. In particular, as per Fig. 2(i), the generated images can evoke interpretations of depth, patterns, motion, lighting effects, textures, objects and locations, and allow certain moods and emotions to be projected onto the image. Through a number of experiments, we produced around 1,000 high quality images using this method, having hand curated them from around 10,000. To further increase the variety and number of the images produced, we experimented with transformations to the lines before they were rendered on the image. We experimented with 6 transformations (diamond, grid, kaleidoscope, necklace, polar and tron), which moved the start and end points of the lines before rendering them. For instance, with the polar transformation, the Cartesian (x, y) coordinates of the line start and end points were transformed to polar coordinates (θ, r). This method produces images of a more circular nature. Some transforms had parameters which were also experimented with, e.g., the number of beads in the necklace transformation. We also trialled a transform which replaced the line with an oval with focal points at the line ends. Example images produced using the different transformation are given in Fig. 2(ii).

As each transformation produces quite visually different images to each other, we increased the volume of images in the archive by taking the 1,000 curated images and applying the transformation before re-rendering them. However, some transforms spread out the lines on the canvas, producing largely empty, unusable, images. Other transformations took quite busy original images and made them too noisy. In particular, the oval transformation produced a high proportion of visually disorienting and messy images, and we had to remove from the dataset around 60% of these images, as they did not provide enough coherence for a reasonable interpretation along any of the lines discussed above, so would not be useful for the Camera Obscurer app. Of the other transformations, around 1 in 20 curated images had to be discarded after the transformation. The necklace and kaleidoscope transforms were most successful aesthetically, as their structure provides balance and symmetry not guaranteed by the other transformations.

We supplemented the 7,000 images which were not removed with 2,500 more, produced in a different run, with no transformation (examples in Fig. 2(i)). However, in preliminary experiments, we found the border around these images was a substantial visual feature, which meant they were rarely retrieved. Hence, we cropped the border before the visual analysis used for indexing them, but kept the full image available for retrieval. In total, we supplied 9,519 abstract art images to Camera Obscurer. In addition, we supplied 20,000 images obtained by passing 20 images representative of diverse aspects of life through 1,000 image filters. As described in [11], the filters were themselves generated via a directed search over a space of trees which pass source images through single transformations like blurring, and also composite two images together, e.g., through a bitwise operator. Many filters produce artistic effects resembling watercolours, embossing, etc., and others produce highly abstracted images. A selection of seven filters applied to two source images is given in Fig. 3. The filters introduce abstraction, new colours, textures and artefacts. For the experiments described below, we also supplied three more datasets, namely: (a) 10,000 images from the *WikiArt* archive (wikiart.org) (b) 10,000 images from the photographic *Archive of the Planet* (http://collections.albert-kahn.hauts-de-seine.fr) and (c) 5,000 thumbnails depicting five colours taken from palettes given in [12].

2.2 Visual Similarity Image Retrieval

Content-based image retrieval is a well studied problem in computer vision, with retrieval problems generally divided into two groups: category-level retrieval [13] and instance-level retrieval [14]. Given a query image of the Sydney Harbour bridge, for instance, category-level retrieval aims to find *any* bridge in a given dataset of images, whilst instance-level retrieval must find *the* Sydney Harbour bridge to be considered a match. Instance retrieval is therefore the much harder problem, and because it has many more applied use cases, it's also the more studied problem.

Traditionally, instance retrieval has been performed using the prominent Bag-of-Words (BoW) model combined with a set of local descriptors, such as scale-invariant feature transforms (SIFT) [15]. Other popular approaches use local descriptor and vector aggregation methods such as the Fisher Vector [16] and the VLAD representation [17], which have been among the state of the art techniques in this domain. In 2012, Krizhevsky et al. [18] demonstrated significant improvements in classification accuracy on the ImageNet [19] challenge using a convolutional neural network (CNN), exceeding prior results in the field by a large margin. Since then, many deep convolutional neural network (DCNN) architectures have been developed such as VGG, ResNet, and Inception. All of these have pushed forward the state of the art in classification accuracy on the ImageNet challenge. Networks trained on the ImageNet classification tasks have been shown to work very well as off the shelf image extractors [20], and show even better results when fine-tuned to datasets for the task at hand [21]. DCNNs are also commonly used as *backbone* networks in more complex multi-stage architectures (functioning as the base-level image feature extractors) which led to state-of-the-art performance in tasks like object detection [22] and object segmentation [23].

Experiments in applying CNNs in the field of image retrieval have shown promising results [24], in some cases surpassing the performance of traditional methods like SIFT [14] and VLAD [21]. Commercial organisations such as Bing and Pinterest have completely moved away from hand engineered features in their image retrieval pipelines, as described in [25] and [26] respectively. Instead, they have adapted one or more DCNN architectures as a way of creating compressed image representations. The evaluation of these systems primarily consists of applying them to datasets with real-life photographs containing easily recognisable objects and/or scenes, with concrete shapes and form, and with inherently natural colour illumination. In contrast, the kinds of computer generated abstract art datasets that we are concerned with, exhibit very few of these naturalistic properties, and differ significantly in terms of their characteristics.

Research in the field of image retrieval, both traditional and CNN based, has not often included evaluating retrieval over abstract images. More diverse datasets have been used in tasks around identifying art genres [27], style [28], and predicting authors of an artistic work [29], but these are in the domain of classification rather than retrieval. Seguin et al. [30] evaluated image retrieval of artworks in the historical period from 1400 to 1800, and showed that BoW with SIFT descriptors didn't transfer well to paintings. They showed that a pre-trained VGG performed well, but they achieved their best results by fine-tuning VGG on their dataset. Performance of VGG on capturing style and content has been well demonstrated by Gatys et al. [31], demonstrating that a pre-trained VGG on a non-domain specific dataset can capture features which identify style and content of artworks.

When implementing retrieval systems for more practical purposes, accuracy is not the only dimension worth measuring. In addition, aspects such as search efficiency, open-source availability, and device support are major considerations. Apple and Google have optimised both their mobile operating systems to natively support running of complex neural network models, offering Core ML and ML Kit respectively. This level of native support combined with the larger open source community around deep learning played a major role in our choice of only evaluating DCNNs for the feature extraction abilities in the Camera Obscurer app. We selected architectures from the three families: VGG-16, ResNet-50, and Inception-v4. All of these networks were downloaded with pre-trained ImageNet weights and used as off-the-shelf feature extractors. For each network, the last non fully-connected layer of the network was used as the feature representation, as it tends to provide a good starting point [32]. We discard the fully-connected layer(s) as they are ImageNet specific and not relevant for feature representation. This meant that for a single image, VGG-16 and ResNet-50 had 2048 features, and Inception-v4 contained 1536 extracted features.

We carried out an empirical evaluation to compare how each network performed on visually similar image retrieval. A sample of the WikiArt corpus was chosen for the evaluation, as it contains a range of different image styles.

Fig. 4. Examples from the 3 networks for the given seed images (leftmost). In general, ResNet-50 (2nd column) performs much better than both VGG-16 (3rd column) and Inception-v4 (4th column) at visual similarity image retrieval for abstract image seeds.

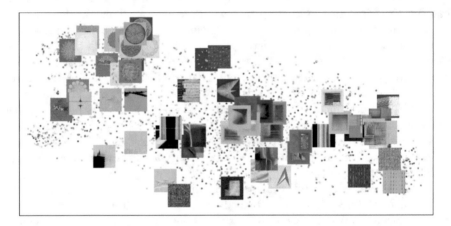

Fig. 5. T-SNE plot showing clusters of visually similar images. The plot shows that a ResNet-50 functions well as a generic feature extractor for highly abstract images.

The dataset contained 10,000 artworks from differing style categories to provide a mix: Realism (2000), Pointillism (500), Contemporary Realism (300), New Realism (200), Impressionism (2000), and Abstract Expressionism (2000). All three networks performed well at retrieval tasks that related to realistic paintings, as expected, but ResNet-50 performed best with respect to more abstract images, as portrayed in Fig. 4, which shows favourable examples from all three networks. In a subjective visual inspection, we found ResNet-50 to be the network that reliably returned the most coherent results, i.e., images that were clearly more visually similar and sensible than those returned by the other networks. In terms of search efficiency, ResNet-50 is also the quickest for inference, requiring in the order of 4 times fewer operations than VGG-16. We were able to run feature extraction using the network at a rate of 15 images per second on an iPhone X using Apple's Core ML toolchain, which is sufficient for our use-case.

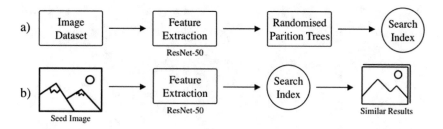

Fig. 6. Architecture of the image retrieval system.

As the WikiArt dataset contained no computer generated abstract images, we further evaluated the performance of ResNet-50 on the dataset containing 9,519 completely abstract images described above. As it would be difficult to construct an objective test-set for evaluating retrieval on these images, we instead evaluate the network extracted features. Figure 5 shows a 2-dimensional projection of a subset of the features (3000 samples) constructed using T-SNE (a popular dimensionality reduction technique). For visual clarity, only 2% of the sample images are shown on the plot, which displays clear image clusters with similar elements: images with circular aspects towards the top left, with matrix-like patterns towards the right. T-SNE doesn't preserve any kind of distance measure, thus no assessments can be made regarding which two groups are the furthest in embedding space. The figure clearly indicates that the ResNet-50 neural network can function as a generic feature extractor even for the most abstract images. Whilst DCNNs may not be the best at representing abstract art for deeper comparisons [33], their attributes such as the off-the-shelf nature, fast inference, and automatic embedding made them a valid choice for Camera Obscurer.

With this embedding representation of an image, the next step in image retrieval is to find and retrieve similar search results. Computing exact nearest neighbours is feasible in evaluation datasets, but as the number of images increases to the high hundreds of thousands, it becomes infeasible to guarantee exact search, especially when search efficiency is a priority. Spatial indexes like k-d trees try to address the problem of exhaustive search by using binary trees to divide up the embedding space, but they only work for feature vectors containing 10s of dimensions, not hundreds or thousands. To address this curse of dimensionality, we turn to *Approximate* Nearest Neighbour (ANN) search algorithms. Methods such as Locality Sensitive Hashing (LSH), Randomised Partition Trees (RPT) don't guarantee exact nearest neighbours, but do promise extremely fast results in very close neighbourhoods of the query.

We used an implementation of RPT combined with random projections that has shown to have competitive performance for both retrieval accuracy and memory usage [34], and is used by the popular music streaming service Spotify for music recommendations (see https://github.com/spotify/annoy). Cosine distance (proportional to euclidean distance of normalised vectors) was applied as the metric when constructing the search index as it is insensitive to vector

Fig. 7. Screenshots from the Camera Obscurer app.

magnitude. The initial search index is computed off-device as it is computation-ally expensive, but subsequent search queries (on an index containing 50,000 images) can be performed on the device in less than 50ms. The query time scales logarithmically as the underlying spatial index is constructed using binary search trees, so the search performance isn't impacted too heavily as the dataset grows. Further improvements to the retrieval accuracy were made by using a forest of RPTs, which incurs minimal cost in search efficiency as queries can be executed in parallel, but increases the likelihood of an exact match, as described in [35].

The architecture of our image retrieval system is shown in Fig. 6. The top half shows the construction of the search index from an image dataset using a ResNet-50 feature extractor combined with RPT based ANN search index. The bottom half portrays how subsequent search queries are computed. Our implementation is sufficiency modular, so that both the feature extractor and the search method can be updated when needed. However, with our current choices, we achieve practically real-time performance on a mobile device.

2.3   User Interface

The Camera Obscurer app is still an early prototype, developed initially for handheld iOS devices. It has a minimal interface consisting of a large video box at the top of the screen – which streams a feed live from the camera on the back of the device – and 10 image boxes below this which show images retrieved from the datasets. It also has a capture button at the bottom of the screen which initiates the process of taking the current still from the video feed as the seed for the visual similarity search. A user can also tap the video box to start the retrieval process, or tap any of the 10 retrieved images, in which case the app uses that as the seed image for a new search. The user can further double tap on any box (image or video) to pin the shown image to a Pinterest board. They can also shake the app in order to replace the currently retrieved images with the previous set. Screenshots from the Camera Obscurer app are given in Fig. 7, highlighting the kinds of image retrieval performed for given seed images taken from local surroundings, both indoors and outdoors.

3   Evaluation

In the case of a standard visual similarity search within a comprehensive image database, the method of evaluating the search algorithm would normally involve pixel-to-pixel matching between the seed image and the retrieved image [36]. In our case, however, the image database is intentionally restricted as the retrieved images are intended to serve as a source of visual inspiration, rather than provide an exact match. For this reason, pixel-to-pixel matching is not a suitable measure of success. Moreover, in our evaluation of Camera Obscurer, we wish to investigate two hypotheses: not only that the retrieved images bear some perceived visual similarity to the seed image, but in addition, that the retrieved images can serve as a source of visual inspiration in the design process. The experiment and pilot study below address these aspects respectively.

3.1   A Visual Similarity Experiment

To measure the perceived visual similarity between a seed image and a retrieved image, we devised a crowdsource experiment utilising Amazon's Mechanical Turk service (mturk.com). 100 participants were presented with a seed image (taken from the MIT-Amobe FiveK Dataset [37]) accompanied by two retrieved images, namely: one image retrieved through our visual similarity search, and the other retrieved randomly. Participants were then asked to select the retrieved image with strongest visual similarity to the seed, basing their decision on colour, mood, texture, shape and content. The seed image dataset was chosen as it provides a suitable representation of the style of images we expect to be taken by future users of Camera Obscurer. Each participant was asked to select the most visually similar image for 100 seed images across three datasets. These comprised 32 seed images from each of: (a) the abstract generated images (Fig. 2) (b) the filtered images (Fig. 3) and (c) the WikiArt dataset (wikiart.org). The remaining 4 seed images were accompanied with one exact match to the seed image and one randomly selected image, to serve as control questions.

As shown in Table 1, the intelligently retrieved images had a greater perceived visual similarity than the randomly retrieved images, as participants correctly

Table 1. Summary results across three image datasets, displaying average accuracy, average cosine distance between seed and retrieved image, Pearson correlation coefficient between average cosine distance and count of incorrect selections, and the percentage population the datasets in each accuracy quartile.

Dataset	Avg Acc.	Avg Dist.	Pearson coefficient	Q1(%)	Q2(%)	Q3(%)	Q4(%)
WikiArt	90.91%	0.423	0.23	25.0	25.0	20.8	62.5
Filtered	85.56%	0.477	0.17	37.5	41.7	33.3	20.8
Abstract	84.89%	0.568	0.17	37.5	33.3	45.9	16.7
All	87.12%	0.490	0.22	100	100	100	100

selected the intelligently retrieved image with an accuracy of 87.12%, averaged over the 96 (non-control) seed images. Using a binomial test, we were able to reject the null hypothesis that participants were as likely to select the random image (as being the closest visually to the seed) as they were to select the intelligently retrieved one ($p < 0.001$). The dataset with the highest accuracy was WikiArt, with the intelligently retrieved images being selected 90.91% of the time. The abstract generated images and filtered image datasets achieved 84.88% and 85.55% accuracy respectively. The average time of appraisal among participants was 8.71 seconds per image. To investigate labelling accuracy per dataset, we present the distribution of images in each dataset across accuracy quartiles in Table 1. We see that WikiArt images account for 62.5% of the images with highest perceived visual similarity (Q4), whereas the filtered and abstract images combined account for 75% of the images with the lowest perceived visual similarity (Q1 and Q2).

The cosine distance between seed and retrieved image appears to be mildly positively correlated ($r = 0.22$) with the number of incorrect selections made, as per the Pearson product moment coefficients in Table 1. The correlation is higher for the WikiArt images, and lower for the filtered and abstract images. This was to be expected given the higher abstraction levels exhibited in the abstract art and filtered images. For illustrative purposes, in Table 2, three example questions from the crowdsourcing test are given, along with the labeling accuracy and the cosine distance. As expected, as the distance of the retrieved image from the seed increases, the crowdsourced labelling accuracy reduces.

Table 2. Example images from each dataset, with accuracy and distance values.

Dataset	Seed	Retrieved	Random	Accuracy	Distance
WikiArt				100.0%	0.452
Filtered				92.86%	0.470
Abstract				32.65%	0.534

3.2   A Design Inspiration Pilot Study

The crowdsourcing results above indicate that abstract images retrieved by the Camera Obscurer app share enough visual similarity with the seed images to be easily distinguishable from images retrieved at random. We felt that this was a sufficient basis on which to carry out a pilot study into the affordances of the app with respect to visual inspiration. To this end, we asked 8 participants to undertake a 3 stage session, for us to gather initial results about app usage.

In stage 1, participants were given a mock design task which involves designing an album cover for an upcoming ambient electronic musician, whose music describes "scenes of sprawling concrete structures juxtaposed with vivid internal landscapes of the creatures inhabiting them". In stage 2, participants were given up to 30 min to employ Camera Obscurer to take photos from their surroundings with which to retrieve images from the five datasets. They were asked to pin any image (from the camera and/or the retrieved images) that they felt provided inspiration for the design task to a Pinterest board, but were given the guidance of pinning none, if they saw fit. Participants were told that they could tap the retrieved images to use them as a seed, and could shake the device to re-load previously retrieved images. In stage 3, participants were interviewed about their proposed design, and asked specifically how images pinned to their board inspired their final design idea, if at all.

Table 3. Summary results for the design inspiration sessions.

Part	Duration	Ext. Seeds	Int. Seeds	Pins	Yield	Abs.	Arch.	Cam.	Fil.	Pal.	WikiArt
1	5m41s	19	5	20	0.83	2	3	10	4	0	1
2	17m56s	42	13	31	0.56	5	3	13	3	3	4
3	12m32s	39	7	34	0.74	4	2	17	5	0	6
4	5m36s	2	15	16	0.94	3	2	1	7	0	3
5	17m05s	30	5	21	0.60	5	1	5	3	1	6
6	20m37s	29	14	59	1.37	17	8	7	9	2	16
7	10m13s	15	52	21	0.31	3	3	0	9	0	6
8	8m53s	11	2	14	1.08	1	3	3	3	1	3
Avg	12m19s	23.38	14.13	27.00	0.72	5.00	3.13	7.00	5.38	0.88	5.63

Summary results from the 8 sessions with (Part)icipants are given in Table 3. We see that the average duration of the sessions was around 12 min, and that, on average in each session, 23 images from (Ext)ernal surroundings were used as seeds, while 14 retrieved (Int)ernal images were used as the seed. Participants pinned 27 images on average per session, and we calculated a yield as the number of images pinned per retrieval action, with Table 3 showing that around 3 images were pinned for every 4 retrievals. The breakdown of pinned images per archive shows that images taken directly from the (Cam)era were the most likely to be pinned (on average 7 per session), with images from the (Abs)tract art, (Fil)tered images and WikiArt equally likely to be pinned next, roughly (around 5 per session). Images from the (Arch)ive of the Planet were more rarely pinned (3 per session), and the colour (Pal)ette images were very unpopular, with less than one on average being pinned per session. Participants 4 and 7 were outliers in the sense of using the internal (retrieved) images more often than the external images as seeds.

In the stage 3 interviews, we endeavoured to determine if the retrieved images had in any way provided inspiration for the design project that participants believed they would not have otherwise gained. One of the aims of the pilot study is to begin to determine how we might measure visual inspiration more concretely, hence we were broad in our assessment of inspiration. We also asked whether any retrievals were unexpected, and if this effected their ability to generate ideas.

A number of trends emerged from the questioning. Firstly, most participants expressed some frustration with the app design, in particular that they couldn't view the images that they had already pinned. We found that if participants started out with a specific design idea in mind, the app could be disruptive to their creative process, particularly for those with a background in design. However, participants that had no previous design experience found that the app actually allowed them to form their design intention. Many participants describe a 'moment' at which a retrieved image sparked an idea for their design. A number of participants enjoyed the abstract nature of the retrieved images, rather than when the app returned a "literal match", e.g., a participant stated that "when taking a photo of grass, and was returned a photo of grass, I didn't really find it interesting". In addition to design inspiration, one participant expressed that the app helps to "capture unconscious feelings" associated to their surroundings.

4   Conclusions and Future Work

We have implemented and tested an approach which has the potential to provide instant visual inspiration when desired in graphic design tasks. The approach involves image retrieval based on visual similarity, and supplies users with abstract art pieces, filtered images, colour palettes, historical photographs and images of human artwork in response to an image taken from their surroundings. We performed a crowdsourcing experiment to test whether the image retrieval performs adequately when applied to the retrieval of abstract images, and we found that this was the case. In addition, we undertook a pilot study into the value of the app in a design scenario, to explore whether it had the potential to provide visual inspiration. We found that – even though it is an early prototype – people were able to employ it in their process with good effect, as on occasion it gave them ideas they probably wouldn't have thought of otherwise.

From the pilot study, we can put forward a number of hypotheses to be tested with a larger empirical study: (i) users pin more images for later reference directly from the camera than retrieved from any dataset (ii) abstracted retrieved images, whether of human art or generative art, are pinned more often for design inspiration than images from reality (iii) graphic design tasks can be influenced by abstract artistic images retrieved in response to visual stimuli, through visual inspiration afforded by the app. To test these hypotheses, we will need to define a more concrete notion of what visual inspiration is in a design setting. Our current thinking is that this needs to be recorded at the time of inspiration, and we expect to implement voice recording as both a design tool for users (i.e., to record their thoughts while they use the app), and as an experimental tool which will enable us to pinpoint moments of visual inspiration.

We also plan to improve the app to take onboard some of the criticisms which arose during the pilot study. Our aim is to keep the interface as simple as possible, while allowing users more requested functionality, such as viewing the images that they have already pinned. One interesting suggestion that we plan to explore is to enable users to choose image datasets in advance and/or upload their own archive for a particular task. In this way, when using Camera Obscurer, say, for a fashion design task, they could upload datasets of garments, fabrics, etc. As indexing the images can be done very quickly, this functionality is definitely a possibility. Noting also that the retrieved abstract art images can be altered/evolved and filtered images have filters which can be altered and employed too, we plan to enable Camera Obscurer to be part of a design workflow where processes are retrieved for in-app usage, rather than just static images.

The work described here is related somewhat to that in [38], where images were evolved specifically to be multi-stable computationally, i.e., ambiguous to machine vision systems, hence analagous to human interpretation of ambiguous imagery. Our approach is also to supply visually ambiguous images to users, but ones which have some visual similarity with a seed image. The ambiguity is a side-effect of matching abstract artistic images to photo-realistic images, and serves as the basis for visual inspiration. Capitalising on such ambiguity, we plan to use similar retrieval methods as above to find ImageNet [19] images with abstract art images as the seed. In this way, Camera Obscurer may be able to provide invented texts to accompany the abstract images it retrieves, e.g., for the centre image at the bottom of Table 2, it might find a match with images of birds in ImageNet, and invent the tag: "red bird in flight", perhaps employing some natural language processing technologies to finesse the output. We plan to test whether this leads to further design inspiration, and in general to explore ways in which AI systems can work as inspirational partners alongside designers. We hope to show that generated abstract art can stimulate design thinking, providing much inspiration for artists and designers.

Acknowledgements. We would like to thank the participants in the pilot study for their time and energy, members of SensiLab for their very useful feedback on the Camera Obscurer app, and the anonymous reviewers for their helpful comments.

References

1. Romero, J., Machado, P. (eds.): The Art of Artificial Evolution: A Handbook on Evolutionary Art and Music. Springer, New York (2008)
2. Elgammal, A., Liu, B., Elhoseiny, M., Mazzone, M.: CAN: creative adversarial networks generating "art" by learning about styles and deviating from style norms. In: Proceedings of the 8th International Conference on Computational Creativity (2017)
3. Reas, C., McWilliams, C.: LUST: Form+Code in Design, Art and Architecture. Princeton Architectural Press, New York (2010)
4. Colton, S.: Evolving a library of artistic scene descriptors. In: Machado, P., Romero, J., Carballal, A. (eds.) EvoMUSART 2012. LNCS, vol. 7247, pp. 35–47. Springer, Heidelberg (2012). https://doi.org/10.1007/978-3-642-29142-5_4
5. Sims, K.: Artificial evolution for computer graphics. Comput. Graph. **25**(4), 319–328 (1991)
6. Todd, S., Latham, W.: Evolutionary Art and Computers. Academic Press, San Diego (1992)
7. McCormack, J.: Aesthetic evolution of L-systems revisited. In: Raidl, G.R., Cagnoni, S., Branke, J., Corne, D.W., Drechsler, R., Jin, Y., Johnson, C.G., Machado, P., Marchiori, E., Rothlauf, F., Smith, G.D., Squillero, G. (eds.) EvoWorkshops 2004. LNCS, vol. 3005, pp. 477–488. Springer, Heidelberg (2004). https://doi.org/10.1007/978-3-540-24653-4_49
8. Correia, J., Machado, P., Romero, J., Carballal, A.: Evolving figurative images using expression-based evolutionary art. In: Proceedings of the ICCC (2013)
9. Machado, P., Correia, J., Romero, J.: Expression-based evolution of faces. In: Machado, P., Romero, J., Carballal, A. (eds.) EvoMUSART 2012. LNCS, vol. 7247, pp. 187–198. Springer, Heidelberg (2012). https://doi.org/10.1007/978-3-642-29142-5_17
10. Colton, S., Cook, M., Raad, A.: Ludic considerations of tablet-based evo-art. In: Di Chio, C., Brabazon, A., Di Caro, G.A., Drechsler, R., Farooq, M., Grahl, J., Greenfield, G., Prins, C., Romero, J., Squillero, G., Tarantino, E., Tettamanzi, A.G.B., Urquhart, N., Uyar, A.Ş. (eds.) EvoApplications 2011. LNCS, vol. 6625, pp. 223–233. Springer, Heidelberg (2011). https://doi.org/10.1007/978-3-642-20520-0_23
11. Colton, S., Torres, P.: Evolving approximate image filters. In: Giacobini, M., Brabazon, A., Cagnoni, S., Di Caro, G.A., Ekárt, A., Esparcia-Alcázar, A.I., Farooq, M., Fink, A., Machado, P. (eds.) EvoWorkshops 2009. LNCS, vol. 5484, pp. 467–477. Springer, Heidelberg (2009). https://doi.org/10.1007/978-3-642-01129-0_53
12. Krause, J.: Color Index. David and Charles (2002)
13. Sharma, G., Schiele, B.: Scalable nonlinear embeddings for semantic category-based image retrieval. In: Proceedings of the IEEE International Conference on Computer Vision (2015)
14. Zheng, L., Yang, Y., Tian, Q.: Sift meets CNN: a decade survey of instance retrieval. IEEE Trans. Pattern Anal. Mach. Intell. **40**(5) (2018)
15. Lowe, D.G.: Distinctive image features from scale-invariant keypoints. Int. J. Comput. Vis. **60**(2), 91–110 (2004)
16. Perronnin, F., Dance, C.: Fisher kernels on visual vocabularies for image categorization. In: Proceedings of the IEEE Conference on Computer Vision and Pattern Recognition (2007)

17. Jegou, H., Perronnin, F., Douze, M., Sánchez, J., Perez, P., Schmid, C.: Aggregating local image descriptors into compact codes. IEEE Trans. Pattern Anal. Mach. Intell. **34**(9), 1704–1716 (2012)
18. Krizhevsky, A., Sutskever, I., Hinton, G.E.: ImageNet classification with deep convolutional neural networks. In: Advances in NIPS (2012)
19. Deng, J., Dong, W., Socher, R., Li, L.J., Li, K., Fei-Fei, L.: ImageNet: a large-scale hierarchical image database. In: Proceedings of the IEEE Computer Vision and Pattern Recognition (2009)
20. Razavian, A.S., Azizpour, H., Sullivan, J., Carlsson, S.: CNN features off-the-shelf: an astounding baseline for recognition. In: Proceedings of the IEEE CVPR Workshops (2014)
21. Azizpour, H., Razavian, A.S., Sullivan, J., Maki, A., Carlsson, S.: From generic to specific deep representations for visual recognition. In: Proceedings of the CVPR Workshops (2015)
22. Lin, T.Y., Goyal, P., Girshick, R., He, K., Dollár, P.: Focal loss for dense object detection. IEEE Trans. Pattern Anal. Mach. Intell. (2018)
23. He, K., Gkioxari, G., Dollár, P., Girshick, R.: Mask R-CNN. In: IEEE ICCV (2017)
24. Babenko, A., Slesarev, A., Chigorin, A., Lempitsky, V.: Neural codes for image retrieval. In: Fleet, D., Pajdla, T., Schiele, B., Tuytelaars, T. (eds.) ECCV 2014. LNCS, vol. 8689, pp. 584–599. Springer, Cham (2014). https://doi.org/10.1007/978-3-319-10590-1_38
25. Hu, H., et al.: Web-scale responsive visual search at Bing. In: Proceedings of the 24th ACM SIGKDD International Conference on Knowledge Discovery and Data Mining (2018)
26. Jing, Y., et al.: Visual search at pinterest. In: Proceedings of the ACM SIGKDD International Conference on Knowledge Discovery and Data Mining (2015)
27. Zujovic, J., Gandy, L., Friedman, S., Pardo, B., Pappas, T.N.: Classifying paintings by artistic genre: an analysis of features and classifiers. In: Proceedings of the IEEE Workshop on Multimedia Signal Processing (2009)
28. Bar, Y., Levy, N., Wolf, L.: Classification of artistic styles using binarized features derived from a deep neural network. In: Agapito, L., Bronstein, M.M., Rother, C. (eds.) ECCV 2014. LNCS, vol. 8925, pp. 71–84. Springer, Cham (2015). https://doi.org/10.1007/978-3-319-16178-5_5
29. Hicsonmez, S., Samet, N., Sener, F., Duygulu, P.: Draw: deep networks for recognizing styles of artists who illustrate children's books. In: Proceedings of the International Conference on Multimedia Retrieval (2017)
30. Seguin, B., Striolo, C., diLenardo, I., Kaplan, F.: Visual link retrieval in a database of paintings. In: Hua, G., Jégou, H. (eds.) ECCV 2016. LNCS, vol. 9913, pp. 753–767. Springer, Cham (2016). https://doi.org/10.1007/978-3-319-46604-0_52
31. Gatys, L.A., Ecker, A.S., Bethge, M.: Image style transfer using convolutional neural networks. In: Proceedings of the IEEE Conference Computer Vision & Pattern Recognition (2016)
32. Razavian, A.S., Sullivan, J., Carlsson, S., Maki, A.: Visual instance retrieval with deep convolutional networks. ITE Trans. Media Tech. Appl. **4**(3), 251–258 (2016)
33. Wang, Y., Takatsuka, M.: SOM based artistic style visualization. In: Proceedings of the IEEE Conference Multimedia and Expo (2013)
34. Aumüller, M., Bernhardsson, E., Faithfull, A.: ANN-benchmarks: a benchmarking tool for approximate nearest neighbor algorithms. In: Beecks, C., Borutta, F., Kröger, P., Seidl, T. (eds.) SISAP 2017. LNCS, vol. 10609, pp. 34–49. Springer, Cham (2017). https://doi.org/10.1007/978-3-319-68474-1_3

35. Dasgupta, S., Sinha, K.: Randomized partition trees for exact nearest neighbor search. In: Proceedings of the Conference on Learning Theory (2013)
36. Keysers, D., Deselaers, T., Ney, H.: Pixel-to-pixel matching for image recognition using hungarian graph matching. In: Rasmussen, C.E., Bülthoff, H.H., Schölkopf, B., Giese, M.A. (eds.) DAGM 2004. LNCS, vol. 3175, pp. 154–162. Springer, Heidelberg (2004). https://doi.org/10.1007/978-3-540-28649-3_19
37. Bychkovsky, V., Paris, S., Chan, E., Durand, F.: Learning photographic global tonal adjustment with a database of input/output image pairs. In: Proceedings of the IEEE Conference on Computer Vision and Pattern Recognition (2011)
38. Machado, P., Vinhas, A., Correia, J., Ekárt, A.: Evolving ambiguous images. In: Proceedings of the IJCAI (2015)

Swarm-Based Identification of Animation Key Points from 2D-medialness Maps

Prashant Aparajeya[1]($\boxtimes$), Frederic Fol Leymarie[2],
and Mohammad Majid al-Rifaie[3]

[1] Headers Ltd., London, UK
p.aparajeya@headers-dev.com
[2] Department of Computing, Goldsmiths, University of London, London, UK
[3] School of Computing and Mathematical Sciences,
University of Greenwich, London, UK

Abstract. In this article we present the use of dispersive flies optimisation (DFO) for swarms of particles active on a medialness map – a 2D field representation of shape informed by perception studies. Optimising swarms activity permits to efficiently identify shape-based keypoints to automatically annotate movement and is capable of producing meaningful qualitative descriptions for animation applications. When taken together as a set, these keypoints represent the full body pose of a character in each processed frame. In addition, such keypoints can be used to embody the notion of the Line of Action (LoA), a well known classic technique from the Disney studios used to capture the overall pose of a character to be fleshed out. Keypoints along a medialness ridge are local peaks which are efficiently localised using DFO driven swarms. DFO is optimised in a way so that it does not need to scan every image pixel and always tend to converge at these peaks. A series of experimental trials on different animation characters in movement sequences confirms the promising performance of the optimiser over a simpler, currently-in-use brute-force approach.

Keywords: Line of Action · Medialness ·
Dispersive flies optimisation · Swarm intelligence · Dominant points ·
Animation

1 Introduction

We consider the problem that faces an animator when drawing a succession of frames to generate the illusion of movement of a character (e.g. Mickey Mouse). A now classic technique, which emerged in the Disney studios in the 1930s is to indicate the main pose via a single curve and redraw and deform this curve to indicate how the character to animate shall change its main allure and position [2]. This technique, often referred to as the "Line of Action" (or LoA) is also used to draw 3D characters, but in this initial work, we focus on using the LoA

© Springer Nature Switzerland AG 2019
A. Ekárt et al. (Eds.): EvoMUSART 2019, LNCS 11453, pp. 69–83, 2019.
https://doi.org/10.1007/978-3-030-16667-0_5

Fig. 1. Examples of the use of the Line of Action to specify the pose of a drawn character. On the left hand side is an example of a rapid sketching of 3D humanoid form using a single main 3D curve indicating the body pose (here from the bottom right foot towards the head, via the hips and torso). To the right is a series of 2D profiles of manikins drawn by Loomis [1] where the different main poses are specified via an LoA.

in conjunction with 2D profiles only (examples in Fig. 1). By simply changing the LoA, e.g. curving or bending it in a different direction, the entire qualitative impression of the drawing can be controlled (Fig. 1). Such a technique has been explored in recent years in computer graphics to provide novel tools in the context of a graphical user interface to map poses from one series of drawn characters to another [3] or to efficiently design from scratch series of poses for 2D or 3D characters [4,5]. Intuitively, there is a close relationship between such a tool used by an expert animator (on paper or via a GUI), and the human perception of shapes in a static pose or in movement. In particular, in the cognitive and vision science literature, a number of experimental results support the idea that medialness provides a powerful substrate to characterise the shape of a profile of an object (articulated or not) [6,7]. Medialness is a mapping of an image of intensity to an image of measures of medialness, i.e. how much a given locus of the 2D image is in relation with object boundaries, as a function of its minimal distance from one or more boundary segments [8]. The larger the amount of simultaneously proximal boundary fragments, the higher the measure of medialness (within a tolerance level). The highest possible medialness value being for the locus at the center of a circle, disk or oval object (refer to Sect. 2 for details).

Medialness tends to create a concentration of information along ridges centred more or less in the middle of a character as well as along the middle of its limbs. In this paper, we propose a new technique to automatically detect a useful set of "hot spots" [9] where medialness is most prominent which can provide a good set of "knobs" or control points to specify an LoA, which the animator can then modify as they see fit (e.g. to change the pose, or to get inspired by an existing drawn set).

The traditional approach to identify loci of interest in medialness is to perform ridge following [10]. This represents a rather exhaustive search procedure usually followed by a thinning to identify a graph-like path in the direction of elongation of a ridge. In our work we replace this path-based approach by an efficient optimisation scheme based on swarms. We explore the application of the recently developed technique for that purpose (refer to Sect. 3). This has the advantage of "zooming-in" to loci of high medialness to get a good sampling of the main ridges without the need for an explicit (graph-like) tracing. A simple interpolation scheme through the identified hot spots provides support of an effective LoA ready for the animator to use.

2   Medialness Measurement

We define the computation of 2D medialness gauge at a point p, modifying Kovács et al. [6] original definition, by adding an orientation constraint such that only those boundary loci which are pointing inward (with respect to the figure) are considered for evaluation (Fig. 2):

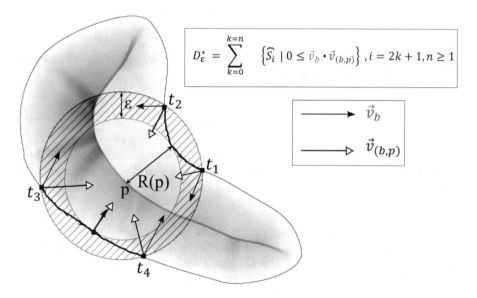

$$D_\varepsilon^* = \sum_{k=0}^{k=n} \left\{ \widehat{S_i} \mid 0 \le \vec{v}_b \cdot \vec{v}_{(b,p)} \right\}, i = 2k+1, n \ge 1$$

$$\longrightarrow \vec{v}_b$$

$$\longrightarrow\!\!\triangleright \vec{v}_{(b,p)}$$

Fig. 2. Illustration of the D_ϵ^* function for a simple shape, defined as an accumulation of boundary segments falling inside an annulus neighborhood of thickness ϵ (shown as darker boundary segments within the annulus' ring) centered around a position p, and such that the associated orientation vector $\vec{v}$ has a dot product with the unit radius vector (taken in the direction of the line from $b(t)$ to p) which is positive. $R(p)$ is taken as the minimum radial distance from p to the nearest contour point.

$$D_\epsilon^+ = \sum_k \widehat{S_{k+1}} \cdot \delta_{v_b \cdot (p-b)} \qquad (1)$$

$$D_\epsilon^- = \sum_k \widehat{S_{k+1}} \cdot \delta_{-v_b \cdot (p-b)} \tag{2}$$

where

$$\delta_x = \begin{cases} 1 & \text{if } x \geq 0 \\ 0 & \text{otherwise} \end{cases} \tag{3}$$

for a point $p = (x_p, y_p)$, vector $b(t) = (x(t), y(t))$ describing (pixel) loci along a piece of 2D bounding contour (B) of the object, and such that v_b is the orientation of the boundary point $b(t)$, $\overrightarrow{v_{(b,p)}}$ is the orientation of the line joining $b(t)$ to p. The positiveness (case of internal medialness, i.e., D_ϵ^+) or negativeness (case of external medialness, i.e., D_ϵ^-) of the scalar product $v_b.(p - b)$ is used to rule out boundary pixels which are oriented away from the given annulus center. We do not consider the geometry (differential continuity) of a contour other than provided by that gradient orientation.

If a boundary pixel has an orientation ϕ, its weight is the value of projection of the square-pixel in the direction of ϕ. Mathematically, this is calculated as:

$$l = \frac{1}{\max(|\sin(\phi)|, |\cos(\phi)|)} \tag{4}$$

The final contribution ∂b of a boundary point b to $\widehat{S_i}$, with orientation ϕ and angular distance θ, is calculated as:

$$\partial b = \begin{cases} l \cos \theta, & \text{if } -\pi/2 \leq \theta \leq \pi/2 \\ 0, & \text{otherwise} \end{cases} \tag{5}$$

The metric $R(p)$, the minimum radial distance to the interior annular shell, is taken as the smallest available distance between p and a bounding contour element:

$$R(p) = \min_t \left\{ |p - b(t)| \mid 0 \leq \overrightarrow{v_b} \cdot \overrightarrow{v_{(b,p)}} \right\} \tag{6}$$

The medialness measure of a point p varies with the two parameters: $R(p)$ and ϵ, where $R(p)$ is the minimum radial distance between p and the bounding contour, and ϵ is the width of the annulus region (capturing object trace or boundary information). Any boundary point b falling inside this annulus that satisfies the definition of Eq. 1 (for interior medialness) or Eq. 2 (for exterior medialness) is added in support for medialness at p. ϵ is selected as a logarithmic function of $R(p)$:

$$\epsilon_p = \kappa \log_b \left(\frac{R(p)}{\kappa} + 1 \right), \, b > 1 \tag{7}$$

Having selected ϵ to be adaptive as a logarithmic function of $R(p)$, the next step is to define a useful logarithmic *base*. We use the same logarithmic base of $e/2$, where e is the Euler's number (approximately 2.718281828), as we report in [11].

3   Dispersive Flies Optimisation

Dispersive flies optimisation (DFO) – first introduced in [12] – is an algorithm inspired by the swarming behaviour of flies hovering over food sources. The swarming behaviour of flies is determined by several factors including the presence of threat which disturbs their convergence on the marker (or the optimum value). Therefore, having considered the formation of the swarms over the marker, the breaking or weakening of the swarms is also noted in the proposed algorithm.

In other words, the swarming behaviour of the flies, in DFO consists of two tightly connected mechanisms, one is the formation of the swarms and the other is its breaking or weakening[1].

Being a population-based continuous optimiser, this algorithm bears similarities with other swarm intelligence algorithms. However, what distinguishes DFO from other population-based optimisers is its sole reliance on agents' position vectors at time t to generate the position vectors for time $t + 1$. Furthermore, other than population size, the algorithm uses a single tunable parameter, Δ, which adjusts the diversity of the population. Therefore, the simplicity of the update equation – due to its lone reliance on the position vector – and having a single parameter to adjust (other than population size) as well as its competitiveness in contrast with other algorithms (as reported later in this section), provides the motivation to use the algorithm for this work. The algorithm and the mathematical formulation of the update equations are introduced next.

The position vectors of the population are defined as:

$$\vec{x}_i^t = \left[x_{i1}^t, x_{i2}^t, ..., x_{iD}^t \right], \qquad i = 1, 2, ..., N \tag{8}$$

where t is the current time step, D is the dimension of the problem space and N is the number of flies (population size). For continuous problems, $x_{id} \in \mathbb{R}$, and in the discrete cases, $x_{id} \in \mathbb{Z}$ (or a subset of $\mathbb{Z}$).

In the first iteration, $t = 0$, the ith vector's d^{th} component is initialised as:

$$x_{id}^0 = x_{\min,d} + u \left(x_{\max,d} - x_{\min,d} \right) \tag{9}$$

where $u = \mathrm{U}(0, 1)$ is the uniform distribution between 0 and 1; $x_{\min}$ and $x_{\max}$ are the lower and upper initialisation bounds of the dth dimension, respectively. Therefore, a population of flies are randomly initialised with a position for each flies in the search space.

On each iteration, the components of the position vectors are independently updated, taking into account the component's value, the corresponding value of

---

[1] Regarding weakening of the population formation, several elements play a role in *disturbing* the swarms of flies; for instance, the presence of a threat causes the swarms to disperse, leaving their current marker; they return to the marker immediately after the threat is over. However, during this period if they discover another marker which matches their criteria closer, they adopt the new marker. Another contributing factor to disturbance is the wind speed, which is suggested to influence the position of the swarm [13].

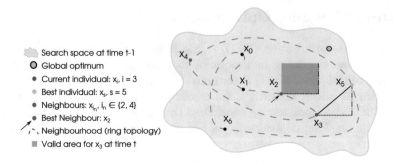

Fig. 3. Sample update of x_i, where $i = 3$ in a 2D space.

the best neighbouring fly with the best fitness (considering ring topology), and the value of the best fly in the whole swarm:

$$x_{id}^t = x_{i_n d}^{t-1} + u(x_{sd}^{t-1} - x_{id}^{t-1}) \tag{10}$$

where $x_{i_n d}^{t-1}$ is the position value of $\vec{x}_i^{t-1}$'s best *neighbouring* fly in the dth dimension at time step $t - 1$, $u = \mathrm{U}(0, 1)$, and x_{sd}^{t-1} is the *swarm's* best fly in the dth dimension at time step $t - 1$. The update equation is illustrated in Fig. 3.

The algorithm is characterised by two main components: a dynamic rule for updating flies position (assisted by a social neighbouring network that informs this update), and communication of the results of the best found fly to other flies.

As stated earlier, the swarm is disturbed for various reasons; one of the impacts of such disturbances is the displacement of flies which may lead to discovering better positions. To consider this eventuality, an element of stochasticity is introduced to the update process. Based on this, individual components of flies' position vectors are reset if a random number, r, generated from a uniform distribution on the unit interval $U(0, 1)$ is less than the *disturbance threshold* or Δ. This guarantees a disturbance to the otherwise permanent stagnation over a likely local minima. Algorithm 1 summarises the DFO algorithm. In this algorithm, each fly is assumed to have two neighbours (i.e. ring topology).

In summary, DFO is a numerical optimiser over continuous or discretised search spaces. DFO is a population based algorithm, originally proposed to search for an optimum value in the feasible solution space. The algorithm's simplicity – in addition to its update equation's bare-bones reliance on only position vectors – comes from having only *one* tunable parameter, Δ, controlling the component-wise dispersion in its population.

As reported in [12], despite the algorithm's simplicity, it is shown that DFO outperforms the standard versions of the well-known Particle Swarm Optimisation [14], Genetic Algorithm (GA) [15] as well as Differential Evolution (DE) [16] on an extended set of benchmarks over three performance measures of error, efficiency and reliability. It was shown that DFO is more efficient in 85% and

---

Algorithm 1. Dispersive flies optimisation

1: **procedure** DFO$(N, D, X_{\min}, X_{\max}, f)$
                                    ▷ INPUT: swarm size, dimensions, lower/upper bounds, fitness func
2:   **for** $i = 0 \to N - 1$ **do**   ▷ Initialisation
3:     **for** $d = 0 \to D - 1$ **do**
4:       $x_{id}^0 \leftarrow U(x_{\min,d}, x_{\max,d})$
5:     **end for**
6:   **end for**
7:   **while** ! termination criteria **do**   ▷ Main DFO loop
8:     **for** $i = 0 \to N - 1$ **do**
9:       $\vec{x}_i$.fitness $\leftarrow f(\vec{x}_i)$
10:    **end for**
11:    $\vec{x}_s = \arg\min [f(\vec{x}_i)], \quad i \in \{0, \ldots, N - 1\}$
12:    **for** $i = 0 \to N - 1$ and $i \neq s$ **do**
13:      $\vec{x}_{i_n} = \arg\min [f(\vec{x}_{(i-1)\%N}), f(\vec{x}_{(i+1)\%N})]$
14:      **for** $d = 0 \to D - 1$ **do**
15:        **if** $U(0, 1) < \Delta$ **then**
16:          $x_{id}^{t+1} \leftarrow U(x_{\min,d}, x_{\max,d})$

17:        **else**
18:          $u \leftarrow U(0, 1)$
19:          $x_{id}^{t+1} \leftarrow x_{i_n d}^t + u(x_{sd}^t - x_{id}^t)$ ▷ Update equation
20:          **if** $x_{id}^{t+1} < x_{\min,d}$ or $x_{id}^{t+1} > x_{\max,d}$ **then**   ▷ Out of bounds
21:            $x_{id}^{t+1} \leftarrow U(x_{\min,d}, x_{\max,d})$
22:          **end if**
23:        **end if**
24:      **end for**
25:    **end for**
26:   **end while**
27:   **return** $\vec{x}_s$
28: **end procedure**

---

more reliable in 90% of the 28 standard optimisation benchmarks used; furthermore, when there exists a statistically significant difference, DFO converges to better solutions in 71% of problem set. Further analysis was also conducted to explore the diversity of the algorithm throughout the optimisation process, a measure that potentially provide more understanding on algorithm's ability to escape local minima. In addition to studies on the algorithm's optimisation performance, DFO has recently been applied to medical imaging [17], optimising machine learning algorithms [18,19], training deep neural networks for false alarm detection in intensive care units [20], computer vision and quantifying symmetrical complexities [21], simulation and gaming [22], analysis of autopoiesis in computational creativity [23].

4   Experiments and Results

In our set of experiments, we have used different frames of animated characters –
as static poses, poses along with the LoA drawn, and frames from an animated (in
movement) sequence. Our target is to identify key-points providing a sampling
of a LoA, in a numerically efficient way. The LoA is itself an abstract concept
(existing in the mind of the professional and trained human animators), and at
this stage in our research we only target a recovery of key-points which can be
interpolated to provide useful if only approximate LoA traces. A future stage
of our research will require a verification experiment of suggested LoA traces
by a group of professionals familiar with the use of the LoA. However, we take
advantage of the fact that, by definition a medialness map helps to localise
features which are approximately (always) near the center lines of an object's
boundary, ensuring that our approximate recovery is at least a potential LoA
trace. In terms of numerical efficiency, a continuum pixel-wise scanning (brute-
force approach) is time consuming when video frames are of high-resolution.
Therefore, we aim to compare DFO with the brute-force search to depict how
DFO is robust and time efficient in finding these key-points.

Fitness Function

For a given image frame, we first extract boundary points and apply the medialness function (defined in Eq. 1) to generate a medialness map. The medialness
values at different positions in this map are taken as the fitness value for the DFO
search process. In other words, medialness serves as the fitness function for DFO.
A peak finding process runs in parallel and assigns as "best flies" those which
are near or at a local medialness peak; other flies, not near local peaks, can also
be initiated (randomly) but with higher fitness values. In following iterations, a
best fly can only be replaced if another fly finds a nearby medialness peak with
higher fitness value, while other flies roam around (refer to Algorithm 1).

   In our experimental set-up, we annotate each character pose with desired
key-points location as (x,y) co-ordinates. This annotation is used to design the
stopping condition for DFO. The DFO search stops if all desired key-points
are found. We categorise our experiments into three parts: (a) standing human
pose, (b) static character, and (c) characters in movement (sequence). For each
of these, we run 100 DFO trials to find best, worst, and average swarming results.
We use 30 flies for swarming where the maximum iteration of convergence is set
to 100. Flies are randomly re-dispersed if either all flies have converged to a peak
or they exceed the maximum allowed number of iterations. These outcomes are
then compared and analysed with the brute approach.

Experiment 1: Standing Human Pose

The first test case is a human silhouette (300 × 520 pixels) in standing pose
containing 18 dominant points, where the key-points resembles to markers found
in MoCAP suits [24,25]. The medialness map (Fig. 4-left) is generated using Eq. 1
and a copy of this map is created to track the flies' converging positions (Fig. 4-
right). Whenever flies converge at a local peak, a circular area of radius ϵ is

removed from the tracking map, which then avoids flies to converge again at the same position. Graph in Fig. 4-right shows the performance of DFO with regards to brute-force approach.

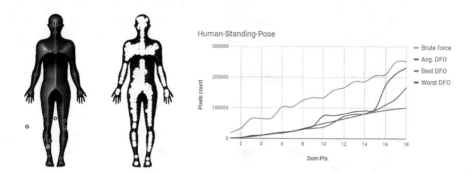

Fig. 4. *Left*: Medialness map of the standing human is represented via grey levels, where high values correspond to brighter shades. Red circles indicate peaks found by the DFO; bright green-circle represents the position of a best-fly, while remaining green circles represent other fly positions. *Middle*: Removal of medialness values in the circular area of radius ϵ (see Eq. 7) for any visited medialness locus. *Right*: graph indicating the number of dominant points found versus the total number of image pixels visited. The yellow curve indicates brute-force approach, while the Green, Red and Blue curves correspond to the results for the Best, Worst and Average DFO search respectively.

DFO converges in a few iterations when the local peaks have high values, but it takes respectively longer time for lower-valued or isolated peaks. From Fig. 4, we can observe that local medialness peaks near head, chest, and abdomen have substantially high medialness values and are surrounded by decreasing medialness field with larger spread area. This allows DFO flies to climb and converge quickly at the peak. We also observe that the DFO flies are less prone to go towards hands and legs area, since the local peaks have lower medialness values with smaller medialness spread. However, smaller peaks are found either after sufficient removal of high valued areas from the tracking map or if flies go in those area because of randomness (in re-dispersal of flies).

Experiment 2: Static Characters Posing

To create a believable (intention of) movement, deciding on the pose of a character is a critical task. The basic principles of to identify a (good) pose are based on the selection of the LoA, a corresponding silhouette, and a sense of weight and balance for the character. Among these overall features, the LoA plays a critical role. By simply changing the LoA – making it more curved, sloped or arched in a different direction – the entire essence of the drawing can be changed. Our proposed shape representation via medial nodes can be seen as a potential psychologically motivated support for LoA [7,11].

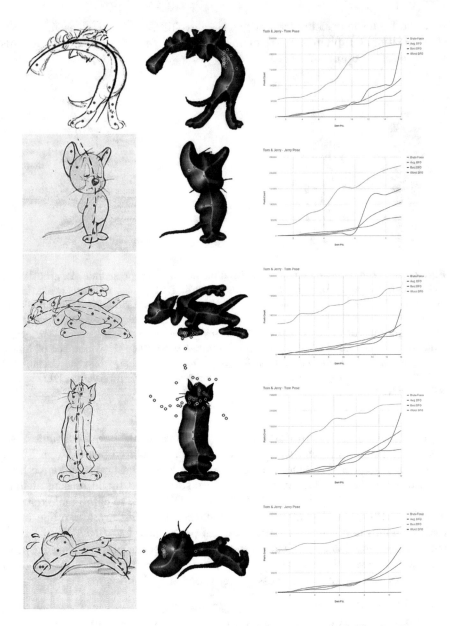

Fig. 5. Tom and Jerry cartoon characters. On the left hand side are different poses of Tom & Jerry drawn on LoA registered with annotated key-points as blue dots. In the middle, medialness maps are represented as gray regions, red circles are peaks found by DFO, bright green circle is the position of best fly, and rest of green circle represent other flies' positions. To the right are time comparison graphs – total number of peaks found versus total number of pixels visited, where Yellow curve indicates brute-force approach, while Green, Red and Blue curves indicate best, worst and average DFO search respectively.

In this experiment set, we aim to identify such medial nodes on the famous "Tom and Jerry" cartoon characters by using DFO. The left part of Fig. 5 is the image frames with original LoAs and registered annotated key-points. It can be seen more clearly how some of these key-points are will fitted along the hand-drawn LoA. In the middle of this figure, a snapshot of DFO processing is shown, where red dots indicated peaks found by DFO and green ones are the swarming flies. Similar to experiment 1, flies converges quickly near the high medialness peak with larger surroundings while lower peaks take slightly longer time. Another interesting find in the graph shown in Fig. 5 is that finding more than 60% of dominant points takes almost similar number of pixel scanning in 100 trials, i.e., the best, worst and average DFO processing is similar in first 60% cases. The graph also indicates that average DFO processing outperforms the brute-force approach.

Fig. 6. First two rows: An animated angry cat running poses. The desired and annotated dominant points are indicated as blue dots. For proper visibility, these single point dots are enlarged into circles. Bottom two rows: Medialness maps are indicated as gray regions, red circles are peaks found by DFO, bright green circle is the position of best fly, and rest of green circle represent other flies' positions.

Experiment 3: Characters in Movement

In this experiment, we have created and annotated set (classes) of characters' frames in movements. Due to the nature of moment cycle, inter class variability

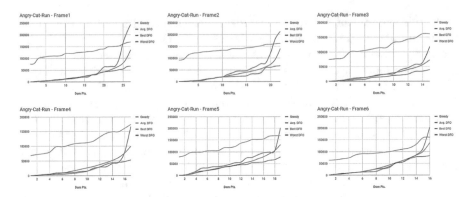

Fig. 7. Time comparison graph, number of peaks found versus total number of image pixel traversed, for frames indicated in Fig. 6 (top rows). Yellow curve indicates brute-force approach, while Green, Red and Blue curves indicate best, worst and average DFO search respectively.

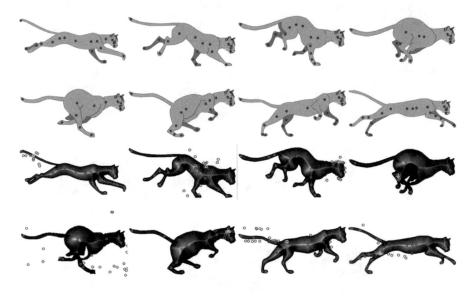

Fig. 8. Animated frames of a running cat. The a priori annotated dominant points are indicated as blue dots. For proper visibility, these single points are enlarged into circles. Bottom two rows: Medialness maps are indicated as gray regions, red circles are peaks found by DFO, while a bright green circle is the position of a current best fly, and other of green circles represent other current fly positions.

exists in both number of frames and desired number of dominant points. A moment cycle consists of the sequence of several body poses where each body pose is different from each other and can have a different number of visible key-points. Example of this can be seen in Figs. 6 (first two rows – "an angry cat" in running poses) & 8 (first two rows – "a normal cat" in running poses). We keep the image resolution the same (500×500) for each class. Snapshots of DFO behaviour are shown in Figs. 6 (bottom two rows) and 8 (bottom two rows).

Similar to our other two experiments, the graph in Fig. 7 shows that more than 75% of keypoints are found by DFO by visiting less than 20–25% of pixels. DFO flies are mostly attracted towards higher and larger medialness areas and the peaks in those areas such as head and abdomen are found more quickly. The tail part of "angry-cat" is much thicker than for a "normal-cat", hence flies converges more quickly.

Our experiments also show that image padding affects the execution time for brute-force approach as it's behaviour is linear. While DFO gets unaffected by such change and produces same search result in similar execution time.

5   Conclusion

Our initial premise was to consider the Line of Action (LoA) for character pose drawing and how the character's pose gets changed by changing the LoA. Our goal is then to design an efficient algorithmic chain to systematically retrieve a useful sampling of "hot spots" in medialness, a generalisation of classic distance maps, Voronoi diagrams and other symmetry-based graph representation. Once a medialness map is produced for existing character boundaries, a brute force approach can be used to seek peaks in magnitude of medialness. For high resolution images or when processing hundreds of frames, a more efficient scheme proves valuable in practice. We have proposed here an efficient search based on Dispersive Flies Optimisation (DFO) which performs well in our targeted scenarios and automatically finds key-points along the LoA quickly and robustly.

Our experimental results indicate that DFO converges much earlier than the currently-in-use brute-force approach and further identifies all dominant points (in labelled test images). Given DFO's GPU-friendly nature, this approach could

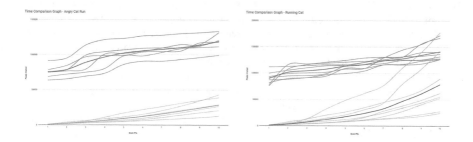

Fig. 9. Time comparison graph to find first 10 dominant points. Light red curves are for brute-force approach for different character frames, while the dark-thick red one indicates the average time execution for the whole movement sequence. Light green curves are for average DFO processing for different frames, while the dark-thick green one is the overall average DFO curve for the full movement sequence. Left graph is for the angry-cat in running pose (see Fig. 6), and right one is for the normal cat in running pose (see Fig. 8)

be further improved for larger, high-resolution input images. For frames where annotation is not available—the usual practical case—a good experimental stopping condition is to limit the swarm search space to up to 70% of the image size. However, in order to find the first 10 best keypoints, our empirical investigation (according to graphs in Fig. 9) suggests that less than 20% of the image size is largely sufficient.

References

1. Loomis, A.: Successful Drawing. Viking Books, New York (1951)
2. Leymarie, F.F., Aparajeya, P.: Medialness and the perception of visual art. Art Percept. **5**(2), 169–232 (2017)
3. Bregler, C., Loeb, L., Chuang, E., Deshpande, H.: Turning to the masters: motion capturing cartoons. ACM Trans. Graph. **21**(3), 399–407 (2002)
4. Guay, M., Cani, M.P., Ronfard, R.: The line of action: an intuitive interface for expressive character posing. ACM Trans. Graph. **32**(6) (2013). Article no. 205
5. Guay, M., Ronfard, R., Gleicher, M., Cani, M.P.: Adding dynamics to sketch-based character animations. In: Proceedings of the Eurographics/ACM Symposium: Expressive Graphics – Sketch-Based Interfaces and Modeling, Istanbul, Turkey, June 2015
6. Kovács, I., Fehér, Á., Julesz, B.: Medial-point description of shape: a representation for action coding and its psychophysical correlates. Vis. Res. **38**(15), 2323–2333 (1998)
7. Leymarie, F.F., Aparajeya, P., MacGillivray, C.: Point-based medialness for movement computing. In: Workshop on Movement and Computing (MOCO), IRCAM, Paris, pp. 31–36. ACM (2014)
8. Burbeck, C.A., Pizer, S.M.: Object representation by cores: identifying and representing primitive spatial regions. Vis. Res. **35**(13), 1917–1930 (1995)
9. Kovács, I.: "Hot spots" and dynamic coordination in Gestalt perception. In: Von der Malsburg, C.V., Phillips, W.A., Singer, W. (eds.) Dynamic Coordination in the Brain: From Neurons to Mind, Chap. 14, pp. 215–28, Strüngmann Forum Reports. MIT Press (2010)
10. Eberly, D., Gardner, R., Morse, B., Pizer, S., Scharlach, C.: Ridges for image analysis. J. Math. Imaging Vis. **4**(4), 353–373 (1994)
11. Aparajeya, P., Leymarie, F.F.: Point-based medialness for 2D shape description and identification. Multimedia Tools Appl. **75**(3), 1667–1699 (2016)
12. al-Rifaie, M.M.: Dispersive flies optimisation. In: Ganzha, M., Maciaszek, L.M.P. (eds.) Proceedings of the 2014 Federated Conference on Computer Science and Information Systems. Annals of Computer Science and Information Systems, vol. 2, pp. 529–538. IEEE (2014). https://doi.org/10.15439/2014F142
13. Downes, J.: The swarming and mating flight of diptera. Ann. Rev. Entomol. **14**(1), 271–298 (1969)
14. Kennedy, J., Eberhart, R.C.: Particle swarm optimization. In: Proceedings of the IEEE International Conference on Neural Networks, vol. IV, pp. 1942–1948. IEEE Service Center, Piscataway (1995)
15. Goldberg, D.E.: Genetic Algorithms in Search. Optimization and Machine Learning. Addison-Wesley Longman Publishing Co. Inc., Boston (1989)
16. Storn, R., Price, K.: Differential evolution - a simple and efficient adaptive scheme for global optimization over continuous spaces (1995). tR-95-012. http://www.icsi.berkeley.edu/~storn/litera.html

17. al-Rifaie, M.M., Aber, A.: Dispersive flies optimisation and medical imaging. In: Fidanova, S. (ed.) Recent Advances in Computational Optimization. SCI, vol. 610, pp. 183–203. Springer, Cham (2016). https://doi.org/10.1007/978-3-319-21133-6_11

18. Alhakbani, H.: Handling class imbalance using swarm intelligence techniques, hybrid data and algorithmic level solutions. Ph.D. thesis, Goldsmiths, University of London, London, United Kingdom (2018)

19. Alhakbani, H.A., al-Rifaie, M.M.: Optimising SVM to classify imbalanced data using dispersive flies optimisation. In: Proceedings of the 2017 Federated Conference on Computer Science and Information Systems, FedCSIS 2017, Prague, Czech Republic, 3–6 September 2017, pp. 399–402. IEEE (2017)

20. Oroojeni, H., al-Rifaie, M.M., Nicolaou, M.A.: Deep neuroevolution: training deep neural networks for false alarm detection in intensive care units. In: European Association for Signal Processing (EUSIPCO) 2018, pp. 1157–1161. IEEE (2018)

21. al-Rifaie, M.M., Ursyn, A., Zimmer, R., Javid, M.A.J.: On symmetry, aesthetics and quantifying symmetrical complexity. In: Correia, J., Ciesielski, V., Liapis, A. (eds.) EvoMUSART 2017. LNCS, vol. 10198, pp. 17–32. Springer, Cham (2017). https://doi.org/10.1007/978-3-319-55750-2_2

22. King, M., al Rifaie, M.M.: Building simple non-identical organic structures with dispersive flies optimisation and A* path-finding. In: AISB 2017: Games and AI, pp. 336–340 (2017)

23. al Rifaie, M.M., Leymarie, F.F., Latham, W., Bishop, M.: Swarmic autopoiesis and computational creativity. Connection Sci., 1–19 (2017). https://doi.org/10.1080/09540091.2016.1274960

24. Ceseracciu, E., Sawacha, Z., Cobelli, C.: Comparison of markerless and marker-based motion capture technologies through simultaneous data collection during gait: proof of concept. PloS One **9**(3) (2014)

25. Aparajeya, P., Petresin, V., Leymarie, F.F., Rueger, S.: Movement description and gesture recognition for live media arts. In: 12th European Conference on Visual Media Production, pp. 19–20. ACM (2015)

Paintings, Polygons and Plant Propagation

Misha Paauw[ID] and Daan van den Berg[✉][ID]

Institute for Informatics, University of Amsterdam, Science Park 904,
1098XH Amsterdam, The Netherlands
mishapaauw@gmail.com, d.vandenberg@uva.nl

Abstract. It is possible to approximate artistic images from a limited number of stacked semi-transparent colored polygons. To match the target image as closely as possible, the locations of the vertices, the drawing order of the polygons and the RGBA color values must be optimized for the entire set at once. Because of the vast combinatorial space, the relatively simple constraints and the well-defined objective function, these optimization problems appear to be well suited for nature-inspired optimization algorithms.

In this pioneering study, we start off with sets of randomized polygons and try to find optimal arrangements for several well-known paintings using three iterative optimization algorithms: stochastic hillclimbing, simulated annealing and the plant propagation algorithm. We discuss the performance of the algorithms, relate the found objective values to the polygonal invariants and supply a challenge to the community.

Keywords: Paintings · Polygons · Plant propagation algorithm · Simulated annealing · Stochastic hillclimbing

1 Introduction

Since 2008, Roger Johansson has been using "Genetic Programming" to reproduce famous paintings by optimally arranging a limited set of semi-transparent partially overlapping colored polygons [1]. The polygon constellation is rendered to a bitmap image which is then compared to a target bitmap image, usually a photograph of a famous painting. By iteratively making small mutations in the polygon constellation, the target bitmap is being matched ever closer, resulting in an 'approximate famous painting' of overlapping polygons.

Johansson's program works as follows: a run gets initialized by creating a black canvas with a single random 3-vertex polygon. At every iteration, one or more randomly chosen mutations are applied: adding a new 3-vertex polygon to the constellation, deleting a polygon from the constellation, changing a polygon's RGBA-color, changing a polygon's place in the drawing index, adding a vertex to an existing polygon, removing a vertex from a polygon and moving a vertex to a new location. After one such mutation, the bitmap gets re-rendered from the

© Springer Nature Switzerland AG 2019
A. Ekárt et al. (Eds.): EvoMUSART 2019, LNCS 11453, pp. 84–97, 2019.
https://doi.org/10.1007/978-3-030-16667-0_6

constellation, compared to the target bitmap by calculating the total squared error over each pixel's RGB color channel. If it improves, it is retained, otherwise the change is reverted. We suspect the program is biased towards increasing numbers of polygons and vertices, but Johansson limits the total numbers of vertices per run to 1500, the total number of polygons to 250 and the number of vertices per polygon from 3 to 10. Finally, although a polygon's RGBA-color can take any value in the program, its alpha channel (opaqueness) is restricted to between 30 and 60[1].

Although the related FAQ honestly enough debates whether his algorithm is correctly dubbed 'genetic programming' or might actually be better considered a stochastic hillclimber (we believe it is), the optimization algorist cannot be unsensitive to the unique properties of the problem at hand. The lack of hard constraints, the vastness of the state space and the interdependency of parameters make it a very interesting case for testing (new) optimization algorithms. But exploring ground in this direction might also reveal some interesting properties about the artworks themselves. Are some artworks easier approximated than others? Which algorithms are more suitable for the problem? Are there any invariants to be found across different artworks?

Today, we present a first set of results. Deviating slightly from Johansson's original program, we freeze the numbers of vertices and polygons for each run, but allow the other variables to mutate, producing some first structural insights. We consistently use three well-known algorithms with default parameter settings in order to open up the lanes of scientific discussion, and hopefully create an entry point for other research teams to follow suit.

2   Related Work

Nature-inspired algorithms come in a baffling variety, ranging in their metaphoric inheritance from monkeys, lions and elephants to the inner workings of DNA [3] (for a wide overview, see [4]). There has been some abrasive debate recently about the novelty and applicability of many of these algorithms [5], which is a sometimes painful but necessary process, as the community appears to be in transition from being explorative artists to rigourous scientists. Some nature-inspired algorithms however, such as the Cavity Method [6,7] or simulated annealing [8], have a firm foothold in classical physics that stretches far beyond the metaphor alone, challenging the very relationship between physics and informatics. Others are just practically useful and although many of the industrial designs generated by nature-inspired or evolutionary algorithms will never transcend their digital existence, some heroes in the field make the effort of actually building them out [9–11]. We need more of that.

The relatively recently minted plant propagation algorithm has also demonstrated its practical worth, being deployed for the Traveling Salesman Problem [12,13], the University Course Timetabling Problem [14] and parametric

---

[1] Details were extracted from Johansson's source code [2] wherever necessary.

optimization in a chemical plant [12]. It also shows good performance on bench-mark functions for continuous optimization [12].

The algorithm revolves around the idea that a given strawberry plant off-shoots many runners in close proximity if it is in a good spot, and few runners far away if it is in a bad spot. Applying these principles to a population of candidate solutions ("individuals") provides a good balance between exploitation and exploration of the state space. The algorithm is relatively easily implemented, and does not require the procedures to repair a candidate solution from crossovers that accidentally violated its hard constraints. A possible drawback however, is that desirable features among individuals are less likely to be combined in a single individual, but these details still await experimental and theoretical verification.

Recent history has seen various initiatives on the intersection between nature-inspired algorithms and art, ranging from evolutionary image transition between bitmaps [15] to artistic emergent patterns based on the feeding behavior of sand-bubbler crabs [16] and non-photorealistic rendering of images based on digital ant colonies [17]. One of the most remarkable applications is the interactive online evolutionary platform 'DarwinTunes' [18], in which evaluation by public choice provided the selection pressure on a population of musical phrases that 'mate' and mutate, resulting in surprisingly catchy melodies.

3   Paintings from Polygons

For this study, we used seven 240×180-pixel target bitmaps in portrait or landscape orientation of seven famous paintings (Fig. 1): *Mona Lisa* (1503) by Leonardo da Vinci, *The Starry Night* (1889) by Vincent van Gogh, *The Kiss* (1908) by Gustav Klimt, *Composition with Red, Yellow and Blue* (1930) by Piet Mondriaan, *The Persistance of Memory* (1931) by Salvador Dali, *Convergence* (1952) by Jackson Pollock, and the only known portrait of Leipzig-based composer Johann Sebastian Bach (1746) by Elias Gottlieb Haussman. Together, they span a wide range of ages, countries and artistic styles which makes them suitable for a pioneering study. Target bitmaps come from the public domain, are slightly cropped or rescaled if necessary, and are comprised of 8-bit RGB-pixels.

Every polygon in the pre-rendering constellation has four byte sized RGBA-values: a red, green, blue and an alpha channel for opaqueness ranging from 0 to 255. The total number of vertices $v \in \{20, 100, 300, 500, 700, 1000\}$ is fixed for each run, as is the number of polygons $p = \frac{v}{4}$. All polygons have at least 3 vertices and, at most $\frac{v}{4} + 3$ vertices as the exact distribution of vertices over the polygons can vary during a run. Vertices in a polygon have coordinate values in the range from 0 to the maximum of the respective dimension, which is either 180 or 240. Finally, every polygon has a drawing index: in rendering the bitmap from a constellation, the polygons are drawn one by one, starting from drawing index 0, so a higher indexed polygon can overlap a lower indexed polygon, but not the other way around. The polygon constellation is rendered to an RGB-bitmap by the Python Image Library [19]. This library, and all the programmatic

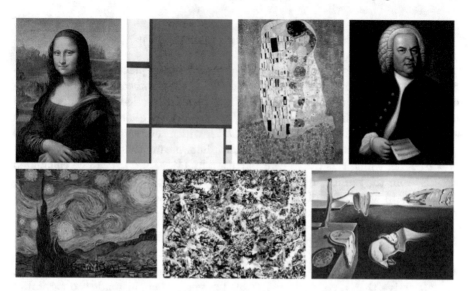

Fig. 1. The paintings used in this study, from a diverse range of countries, eras and artistic styles. From top left, clockwise: *Mona Lisa* (1503, Leonardo da Vinci), *Composition with Red, Yellow, and Blue* (1930, Piet Mondriaan), *The Kiss* (1908, Gustav Klimt), Portrait of J.S. Bach (1746, Elias Gottlieb Hausmann), *The Persistence of Memory* (1931, Salvador Dali), *Convergence* (1952, Jackson Pollock), and *The Starry Night* (1889, Vincent van Gogh).

resources we use, including the algorithms, are programmed in Python 3.6.5 and are publicly available [20].

After rendering, the proximity of the rendered bitmap to the target bitmap, which is the mean squared error (MSE) per RGB-channel, is calculated:

$$\sum_{i=1}^{180 \cdot 240 \cdot 3} \frac{(Rendered_i - Target_i)^2}{180 \cdot 240} \qquad (1)$$

in which $Rendered_i$ is a Red, Green or Blue channel in a pixel of the rendered bitmap's pixel and $Target_i$ is the corresponding channel in the target bitmap's pixel. It follows that the best possible MSE for a rendered bitmap is 0, and is only reached when each pixel of the rendered bitmap is identical to the target bitmap. The worst possible fitness is $255^2 \cdot 3 = 195075$, which corresponds to the situation of a target bitmap containing only 'extreme pixels' with each of the three RGB-color registers at either 0 or 255, and the rendered bitmap having the exact opposite for each corresponding color register.

Although it is a rare luxury to know the exact maximal and minimal objective values (in this case the MSE) for any optimization problem, they're hardly usable due to the sheer vastness of the state space $|S|$ of possible polygon constellations. It takes some special tricks to enable computers to perform even basic operations on numbers of this magnitude, but nonetheless, some useful bounds can be given by

$$S = \alpha \cdot (240 \cdot 180)^v \cdot (256^4)^{\frac{v}{4}} \cdot (\frac{v}{4})! \tag{2}$$

in which $(240 \cdot 180)^v$ is the combination space for vertex placement, $(256^4)^{\frac{v}{4}}$ represents all possible polygon colorings and $(\frac{v}{4})!$ is the number of possible drawing orders. The variable α reflects the number of ways the vertices can be distributed over the polygons. In our case, $\frac{3v}{4}$ vertices are priorly allocated to assert that every polygon has at least three vertices, after which the remaining $\frac{v}{4}$ vertices are randomly distributed over the $\frac{v}{4}$ polygons. This means that for these specifics, $\alpha = P(\frac{v}{4})$, in which P is the integer partition function, and $|S|$ can be calculated exactly from the number of vertices v.

The state space size of the target bitmaps is also known. It is, neglecting symmetry and rotation, given by

$$256^{3^{(180 \cdot 240)}} \approx 7.93 \cdot 10^{312,107} \tag{3}$$

which reflects the three RGB-values for each pixel in the bitmap. From Eq. 2, we can derive that a constellation of 39,328 vertices could render to a total of $\approx 1.81 \cdot 10^{312,109}$ different bitmap images. These numbers are incomprehensibly large, but nonetheless establish an exact numerical lower bound on the number of vertices we require to render every possible bitmap image of 180×240 pixels. A feasible upper bound can be practically inferred: if we assign a single square polygon with four vertices to each pixel, we can create every possible bitmap of aforementioned dimensions with $180 \cdot 240 \cdot 4 = 172,800$ vertices.

Upholding these bounds and assumptions, we would need somewhere between 39,328 en 172,800 vertices to enable the rendering of every possible target bitmap and thereby guarantee the theoretical reachability of a perfect MSE of 0. These numbers are nowhere near practically calculable though, and therefore it can not be said whether the perfect MSE is reachable for vertex numbers lower than 172,800, or even what the best reachable value is. But even if a perfect MSE was reachable, it is unclear what a complete algorithm for this task would look like, let alone whether it can do the job within any reasonable amount of time. What we can do however, is compare the performance of three good heuristic algorithms, which we will describe in the next section.

4   Three Algorithms for Optimal Polygon Arrangement

We use a stochastic hillclimber, simulated annealing, and the newly developed plant propagation algorithm to optimize parameter values for the polygon constellations of the seven paintings (Fig. 2). The mutation operators are identical for all algorithms, as is the initialization procedure:

4.1   Random Initialization

Both the stochastic hillclimber and the simulated annealing start off with a single randomly initialized polygon constellation, whereas the plant propagation algorithm has a population of $M = 30$ randomly initialized constellations (or *individuals*). Initialization first creates a black canvas with the dimensions of the target bitmap. Then, v vertices and $p = \frac{v}{4}$ polygons are created, each with the necessary minimum of three vertices, and the remaining $\frac{v}{4}$ vertices randomly distributed over the polygons. All vertices in all polygons are randomly placed on the canvas with $0 \leq x < Xmax$ and $0 \leq y < Ymax$. Finally, all polygons are assigned the four values for RBGA by randomly choosing values between 0 and 255. The only variables which are not actively initialized are the drawing indices i, but the shrewd reader might have already figured out that this is unnecessary, as all other properties are randomly assigned throughout the initialization process.

4.2   Mutations

All three algorithms hinge critically on the concept of a mutation. To maximize both generality and diversity, we define four types of mutation, all of which are available in each of the three algorithms:

1. **Move Vertex** randomly selects a polygon, and from that polygon a random single vertex. It assigns a new randomly chosen location from $0 \leq x < xMax$ and $0 \leq y < yMax$.
2. **Transfer Vertex:** randomly selects two different polygons $p1$ and $p2$, in which $p1$ is not a triangle. A random vertex is deleted from $p1$ and inserted in $p2$ on a random place between two other vertices of $p2$. Because of this property, the shape of $p2$ does not change, though it might change in later iterations because of operation 1. Note that this operation enables diversification of the polygon types but it keeps the numbers of polygons and vertices constant, facilitating quantitative comparison of the end results.
3. **Change Color:** randomly chooses either the red, green, blue or alpha channel from a randomly chosen polygon and assigns it a new value $0 \leq q \leq 255$.
4. **Change Drawing Index:** randomly selects a polygon and assigns it a new index in the drawing order. If the new index is lower, all subsequent polygons get their index increased by one ("moved to the right"). If the new index is higher, all previous polygons get their index decreased by one ("moved to the left").

It follows that for any constellation at any point, there are v possible move vertex operators, $\frac{v}{4}$ possible transfer vertex operators, $\frac{3v}{4}$ possible change color operators, and $\frac{v}{4}$ possible change drawing index operators. As such, the total number of possible mutation operators for a polygon constellation of v vertices is $\frac{9v}{4}$.

4.3   Stochastic Hillclimber and Simulated Annealing

A run of the stochastic hillclimber works as follows: first, it initializes a random constellation of v vertices and $\frac{v}{4}$ polygons as described in Sect. 4.1. Then, it renders the polygon constellation to a bitmap and calculates its MSE respective to the target bitmap. Each iteration, it selects one of the four mutation types with probability $\frac{1}{4}$, and randomly selects appropriate operands for the mutation type with equal probability. Then it re-renders the bitmap; if the MSE increases (which is *un*desirable), the change is reverted. If not, it proceeds to the next iteration unless the prespecified maximum number of iterations is reached, in which case the algorithm halts.

Simulated annealing [8] works in the exact same way as the stochastic hill-climber, accepting random improvements, but has one important added feature: whenever a random mutation *increases* the error on the rendered bitmap, there is still a chance the mutation gets accepted. This chance is equal to

$$e^{\frac{-\Delta MSE}{T}} \tag{4}$$

in which ΔMSE is the increase in error and T is the 'temperature', a variable depending on the iteration number i. There are many ways of lowering the temperature, but we use the cooling scheme by Geman and Geman [21]:

$$T = \frac{c}{ln(i+1)}. \tag{5}$$

This scheme is the only one proven to be optimal [22], meaning that it is guaranteed to find the global minimum of the MSE as i goes to infinity, given that the constant c is "the highest possible energy barrier to be traversed". To understand what this means, and correctly implement it, we need to examine the roots of simulated annealing.

In condensed matter physics, $e^{\frac{-E}{kT}}$ is known as "Boltzmann's factor". It reflects the chance that a system is in a higher state of energy E, relative to the temperature. In metallurgic annealing, this translates to the chance of dislocated atoms moving to vacant sites in the lattice. If such a move occurs, the system crosses a 'barrier' of higher energy which corresponds to the distance the atom traverses. But whenever the atom reaches a vacant site, the system drops to a lower energetic state, from which it is unlikelier to escape. More importantly: it removes a weakness from the lattice and therefore the process of annealing significantly improves the quality of the metal lattice. The translation from energy state to the MSE in combinatorial optimization shows simulated annealing truly fills the gap between physics and informatics: for the algorithm to escape from a local minimum, the MSE should be allowed to increase, to eventually find a better minimum.

So what's left is to quantify "the highest possible energy barrier to be traversed" for the algorithm to guarantee its optimality. As we have calculated the upper bound for MSE in Sect. 2, we can set $c = 195075$ for our 180×240-paintings, thereby guaranteeing the optimal solution (eventually). The savvy reader might feel some suspicion along the claim of optimality, especially considering the open issue on $P \overset{?}{=} NP$, but there is no real contradiction here. The snag is in i going to infinity; we simply do not have that much time and if we would, we might just as well run a brute-force algorithm, or even just do stochastic sampling to find the global minimum. In infinite time, everything is easy.

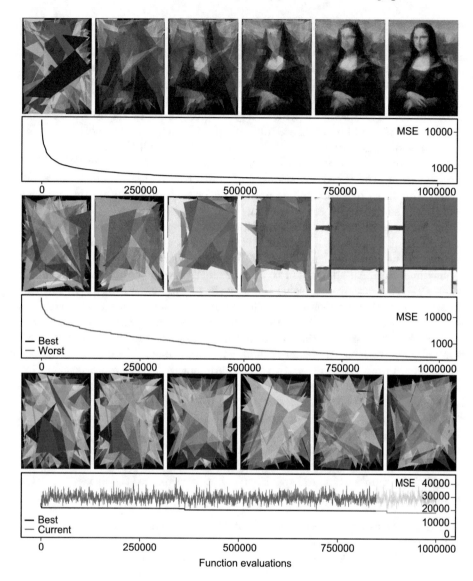

Fig. 2. Typical runs for the stochastic hillclimber (top), the plant propagation algorithm (middle) and simulated annealing (bottom). Polyptychs of rendered bitmaps illustrate the visual improvement from left to right, while the in-between graphs show the corresponding decrease in MSE throughout the iterative improvement process of the respective algorithm.

4.4   Plant Propagation

The plant propagation algorithm (PPA) is a relatively new member of the optimization family and the only population-based algorithm we use. First, fitness values are assigned. Normalized to (0,1) within the current population, this 'relative normalized fitness assignment' might be beneficial for these kinds of optimization problems in which the absolute values of the MSE are (usually)(practically) unknown. Each individual will then produce between 1 and $nMax = 5$ offspring, proportional to its fitness, which will be mutated *inversely* proportional to the normalized (relative) fitness. So fitter individuals produce more offspring with small mutations, and unfitter individuals produce fewer offspring with large mutations. The notions of 'large' and 'small' mutations however, needs some careful consideration.

In the seminal work on PPA, mutations are done on real-valued dimensions of continuous benchmark functions [12], which has the obvious advantage that the size of the mutation can be scaled directly proportional to the (real-valued) input dimension. An adaptation to a non-numerical domain has also been made when PPA was deployed to solve the Traveling Salesman Problem [13]. In this case, the size of the mutation is reflected in the number of successive 2-opts which are applied to the TSP-tour. It does not, however take into account the multiple k-opts might overlap, or how the effect of the mutation might differ from edge to edge, as some edges are longer than others.

So where does it leave us? In this experiment, a largely mutated offspring could either mean 'having many mutations' or 'having large mutations'. Consulting the authors, we tried to stick as closely to the idea of (inverse) proportionality and find middle ground in the developments so far. In the benchmark paper, all dimensions are mutated by about 50% in the worst-case fitness. In the TSP-paper, the largest mutations vary between 7% and 27% of the edges, depending on the instance size. We chose to keep the range of mutations maximal, while assigning the number of mutations inversely proportional to the fitness, resulting in the following PPA-implementation:

1. **Initialization:** create a population of $M = 30$ individuals (randomly initialized polygon constellations).
2. **Assign Fitness Values:** first, calculate the MSE_i for all individuals i in the population, and normalize the fitness for each individual:

$$f_i = \frac{MSE_{max} - MSE_i}{MSE_{max} - MSE_{min}} \tag{6}$$

$$N_i = \frac{1}{2}(tanh(4 \cdot (1 - f_i) - 2) + 1) \tag{7}$$

in which MSE_{max} and MSE_{min} are the maximum and minimum MSE in the population, f_i is an individual's relative fitness and N_i is an individual's normalized relative fitness.
3. **Sort Population** on fitness, keep the fittest M individuals and discard the rest.

4. **Create Offspring:** each individual i in the population creates n_r new individuals as

$$n_r = \lceil n_{max} \cdot N_i \cdot r_1 \rceil \tag{8}$$

with m_r mutations as

$$m_r = \lceil \frac{9v}{4} \cdot \frac{1}{n_{max}} \cdot (1 - N_i) \cdot r_2 \rceil \tag{9}$$

in which $\frac{9v}{4}$ is the number of possible mutation operations for a constellation of v vertices and r_1 and r_2 are random numbers from $(0,1)$.

5. **Return to step 2** unless the predetermined maximum number of function evaluations is exceeded, in which case the algorithm terminates. In the experiments, this number is identical to the number of iterations for the other two algorithms, that both perform exactly one evaluation each iteration.

5   Experiments and Results

We ran the stochastic hillclimber and simulated annealing for 10^6 iterations and completed five runs for each painting and each number of $v \in \{20, 100, 300, 500, 700, 1000\}$. Simulated annealing was parametrized as described in Sect. 4.2 and we continually retained the best candidate solution so far, as the MSE occasionally increases during a run. The plant propagation algorithm was parametrized as stated in Sect. 4.3 and also completed five runs, also until 10^6 function evaluations had occurred. After completing a set of five runs, we averaged the last (and therefore best) values (Fig. 3).

For almost all paintings with all numbers of v, the stochastic hillclimber is by far the superior algorithm. It outperformed the plant propagation algorithm on 48 out of 49 vertex-painting combinations with an average MSE improvement of 766 and a maximum improvement of 3177, on the Jackson Pollock with 1000 vertices. Only for the Mona Lisa with 20 vertices, PPA outperformed the hillclimber by 37 MSE. Plant propagation was the second best algorithm, better than simulated annealing on all painting-vertex combinations by an average improvement of 17666 MSE. The smallest improvement of 4532 MSE was achieved on Bach with 20 vertices and a largest improvement of 46228 MSE on the Mondriaan with 20 vertices. A somewhat surprising observation is that for both the stochastic hillclimber and plant propagation, end results improved with increasing vertex numbers, but only up to about $v = 500$, after which the quality of the rendered bitmaps largely leveled out. This pattern also largely applies to the end results of simulated annealing, with the exception of Bach, which actually got worse with increasing numbers of v. This surprising and somewhat counterintuitive phenomenon might be explained from the fact the algorithm does not perform very well, and the fact that Hausmann's painting is largely black which is the canvas' default color. When initialized with more vertices (and therefore more polygons), a larger area of the black canvas is covered with colored polygons, increasing the default MSE-distance from Bach to black and back.

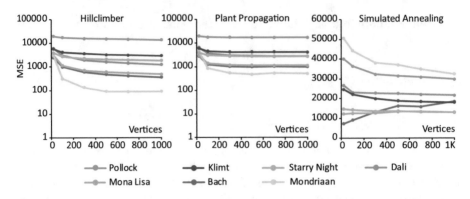

Fig. 3. Best results of five runs for the three algorithms on all seven paintings for all numbers of vertices. Remarkably enough, results often do not improve significantly with numbers of vertices over 500. Note that results for the hillclimber and plant propagation are depicted on a logarithmic scale.

6   Discussion, Future Work, and Potential Applications

In this work, we set out to compare three different algorithms to optimize a constellation of semi-transparent, partially overlapping polygons to approximate a set of famous paintings. It is rather surprising that in this exploration, the simplest algorithm performs best. As the hillclimber is extremely susceptible to getting trapped in local minima, this might indicate the projection of the objective function from the state space to the MSE might be relatively convex when neighbourized with mutation types such as ours. Another explanation might be the presence of many high-quality solutions despite the vastness of the state space.

Plant propagation is not very vulnerable to local minima, but it does not perform very well considering its track record on previous problems. This might be due to the mutation operators, or the algorithm's parametrization, but also to the vastness of the state space; even if the algorithm does avoid local minima, their attractive basins might be so far apart that the maximum distance our mutations are likely to traverse are still (far) too small.

The rather poor performance of simulated annealing can be easily explained from its high temperature. The logarithmic Geman&Geman-scheme in itself cools down rather slowly, even over a run of a million iterations. But our choice of c-value might also be too high. The canvas is not comprised of 'extreme colors', and neither are the target bitmaps, so the maximum energy barrier (increase in MSE) might in practice be much lower than our upper bound of 195075. An easy way forward might be to introduce an 'artificial Boltzmann constant', effectively facilitating faster temperature decrease. This method however, does raise some theoretical objections as the principle of 'cooling down slowly' is a central dogma in both simulated and metallurgic annealing.

An interesting observation can be made in Fig. 3. Some paintings appear to be harder to approximate with a limited number of polygons than others, leading to a 'polygonicity ranking' which is consistently shared between the hillclimber and plant propagation. The essentially grid based work of Mondriaan seems to be more suitable for polygonization than Van Gogh's Starry Starry Night, which consists of more rounded shapes, or Jackson Pollock's work which has no apparent structure at all. To quantitatively assert these observations, it would be interesting to investigate whether a predictive data analytic could be found to *a priori* estimate the approximability of a painting. This however, requires a lot more theoretical foundation and experimental data.

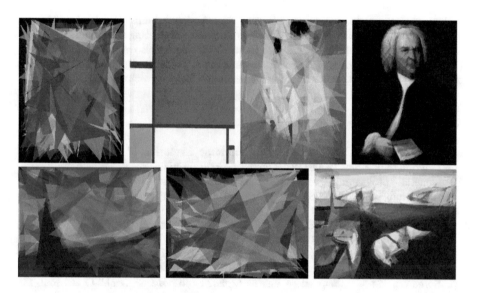

Fig. 4. Best results for various paintings and various algorithms. From top left, clockwise: 1: Mona Lisa, SA, $v = 1000$. 2: Mondriaan, HC, $v = 1000$. 3: Klimt, PPA, $v = 1000$. 4: Bach, HC $v = 1000$. 5: Dali, HC, $v = 1000$. 6: Jackson Pollock, SA, $v = 1000$. 7: Starry Night, PPA, $v = 1000$. This Mondriaan is the best approximation in the entire experiment; when looking closely, even details like irregularities in the painting's canvas and shades from the art gallery's photograph can be made out.

7   Challenge

We hereby challenge all our colleagues in the field of combinatorial optimization to come up with better results than ours (Fig. 4). Students are encouraged to get involved too. Better results include, but are not limited to:

1. Better instance results: finding better MSE-values for values of v and p identical to ours, regardless of the method involved. If better MSE-values are found for smaller v and p, than ours, they are considered stronger.

2. Better algorithmic results: finding algorithms that either significantly improve MSE-results, or achieve similar results in fewer function evaluations.
3. Better parameter settings: finding parameter settings for our algorithms that that either significantly improve MSE-results, or achieve similar results in fewer function evaluations. We expect this is well possible.
4. Finding multiple minima: it is unclear whether any minimum value of MSE can have multiple different polygon constellations, apart from symmetric results.

For resources, as well as improved results, refer to our online page [20]. We intend to keep a high score list.

Acknowledgements. We would like to thank Abdellah Salhi (University of Essex) and Eric Fraga (University College London) for their unrelenting willingness to discuss and explain the plant propagation algorithm. A big thanks also goes to Arnoud Visser (University of Amsterdam) for providing some much-needed computing power towards the end of the project, and to Jelle van Assema for helping with the big numbers.

References

1. Roger Johansson blog: Genetic programming: Evolution of Mona Lisa. https://rogerjohansson.blog/2008/12/07/genetic-programming-evolution-of-mona-lisa/
2. Genetic programming: Mona Lisa source code and binaries. https://rogerjohansson.blog/2008/12/11/genetic-programming-mona-lisa-source-code-and-binaries/
3. Eiben, A.E., Smith, J.E.: Introduction to Evolutionary Computing, 1st edn. Springer, Heidelberg (2003). https://doi.org/10.1007/978-3-662-05094-1
4. Fister Jr., I., Yang, X.S., Fister, I., Brest, J., Fister, D.: A brief review of nature-inspired algorithms for optimization (2013). arXiv preprint: arXiv:1307.4186
5. Sörensen, K.: Metaheuristics—the metaphor exposed. Int. Trans. Oper. Res. **22**, 1–16 (2015)
6. Mézard, M., Parisi, G., Zecchina, R.: Analytic and algorithmic solution of random satisfiability problems. Science **297**(5582), 812–815 (2002)
7. Mézard, M., Parisi, G.: The cavity method at zero temperature. J. Stat. Phys. **111**(1–2), 1–34 (2003)
8. Kirkpatrick, S., Gelatt, C.D., Vecchi, M.: Optimization by simulated annealing. Science **220**(4598), 671–680 (1983)
9. Hornby, G., Globus, A., Linden, D., Lohn, J.: Automated antenna design with evolutionary algorithms. In: Space, p. 7242 (2006)
10. Moshrefi-Torbati, M., Keane, A.J., Elliott, S.J., Brennan, M.J., Rogers, E.: Passive vibration control of a satellite boom structure by geometric optimization using genetic algorithm. J. Sound Vibr. **267**(4), 879–892 (2003)
11. Jelisavcic, M., et al.: Real-world evolution of robot morphologies: a proof of concept. Artif. Life **23**(2), 206–235 (2017)
12. Salhi, A., Fraga, E.: Nature-inspired optimisation approaches and the new plant propagation algorithm. In: Proceeding of the International Conference on Numerical Analysis and Optimization (ICeMATH 2011), Yogyakarta, Indonesia (2011)

13. Selamoğlu, Bİ., Salhi, A.: The plant propagation algorithm for discrete optimisation: the case of the travelling salesman problem. In: Yang, X.-S. (ed.) Nature-Inspired Computation in Engineering. SCI, vol. 637, pp. 43–61. Springer, Cham (2016). https://doi.org/10.1007/978-3-319-30235-5_3

14. Cheraita, M., Haddadi, S., Salhi, A.: Hybridizing plant propagation and local search for uncapacitated exam scheduling problems. Int. J. Serv. Oper. Manag. (in press). http://www.inderscience.com/info/ingeneral/forthcoming.php?jcode=ijsom

15. Neumann, A., Alexander, B., Neumann, F.: Evolutionary image transition using random walks. In: Correia, J., Ciesielski, V., Liapis, A. (eds.) EvoMUSART 2017. LNCS, vol. 10198, pp. 230–245. Springer, Cham (2017). https://doi.org/10.1007/978-3-319-55750-2_16

16. Richter, H.: Visual art inspired by the collective feeding behavior of sand-bubbler crabs. In: Liapis, A., Romero Cardalda, J.J., Ekárt, A. (eds.) EvoMUSART 2018. LNCS, vol. 10783, pp. 1–17. Springer, Cham (2018). https://doi.org/10.1007/978-3-319-77583-8_1

17. Semet, Y., O'Reilly, U.-M., Durand, F.: An interactive artificial ant approach to non-photorealistic rendering. In: Deb, K. (ed.) GECCO 2004, Part I. LNCS, vol. 3102, pp. 188–200. Springer, Heidelberg (2004). https://doi.org/10.1007/978-3-540-24854-5_17

18. MacCallum, R.M., Mauch, M., Burt, A., Leroi, A.M.: Evolution of music by public choice. Proc. Natl. Acad. Sci. **109**(30), 12081–12086 (2012)

19. Python image library 5.1.0. https://pillow.readthedocs.io/en/5.1.x/

20. Heuristieken.nl paintings, polygons, and plant propagation. http://heuristieken.nl/wiki/index.php?title=Paintings_from_Polygons

21. Geman, S., Geman, D.: Stochastic relaxation, Gibbs distributions, and the Bayesian restoration of images. IEEE Trans. Pattern Anal. Mach. Intell. **6**, 721–741 (1984)

22. Nourani, Y., Andresen, B.: A comparison of simulated annealing cooling strategies. J. Phys. A Math. Gen. **31**, 8373–8385 (1998)

Evolutionary Games for Audiovisual Works: Exploring the Demographic Prisoner's Dilemma

Stefano Kalonaris[✉][iD]

RIKEN Center for Advanced Intelligence Project, Tokyo, Japan
stefano.kalonaris@riken.jp

Abstract. This paper presents a minimalist audiovisual display of an evolutionary game known as the Demographic Prisoner's Dilemma, in which cooperation emerges as an evolutionary stable behaviour. Abiding by a dialogical approach foregrounding the dynamical negotiation of the author's aesthetic aspirational levels, the cross-space mapping between the formal model and the audiovisual work is explored, and the system undergoes several variations and modifications. Questions regarding computational measures of beauty are raised and discussed.

Keywords: Evolutionary games · Demographic Prisoner's Dilemma · Computational aesthetics

1 Introduction

Traditional approaches to game theory and decision-making are based on strong assumptions about what agents know, and are conditioned upon a notion of rationality. Evolutionary game theory instead considers the dynamics of behaviour in large, strategically interacting populations. More specifically, population games posit that agents switch strategies only sporadically, and stick to simple rules of behaviour for their decision-making. While evolutionary algorithms have been used in music, art and design, game-theoretical perspectives on population dynamics are rarely considered in these domains. This paper explores evolutionary games and disequilibrium analysis by proposing a simple conceptual blend that maps the emergence of cooperation in a Demographic Prisoner's Dilemma (DPD) to a minimalist audiovisual (AV) work. Several versions of the DPD are explored, by gradually increasing the properties of the agents (e.g., maximum age, mutation of strategy). As this process unfolds, aesthetic considerations rise to the fore, prompting modifications of the cross-space mapping to meet the author's aspirational levels in the AV domain.

2 Context

Evolutionary algorithms (EA) are population-based metaheuristic optimisation methods inspired by biological evolution and are used primarily in the field of

---

The original version of this chapter was revised: Footnote 3 was removed. The correction to this chapter is available at https://doi.org/10.1007/978-3-030-16667-0_17

© Springer Nature Switzerland AG 2019, corrected publication 2021
A. Ekárt et al. (Eds.): EvoMUSART 2019, LNCS 11453, pp. 98–109, 2019.
https://doi.org/10.1007/978-3-030-16667-0_7

artificial intelligence. However, EA are also employed in music [1–5], art [6–9], design [10,11], fashion [12,13] and architecture [14,15]. Game theory is a branch of applied mathematics which studies scenarios of conflict or cooperation and has applications in social and computer science, but has also had marginal applications in music [16–21]. A particular perspective that combines these two theories is evolutionary game theory, which can be employed when trying to model strategic interactions that are difficult to be expressed as dynamic equations, thus requiring numerical simulations with many step iterations. Evolutionary game theoretical approaches in the context of the arts are rare and, to the author's knowledge, there is no precedent of an artistic exploration of the DPD. Before delving into the formal details of the latter, a brief description of evolutionary games is offered.

3 Evolutionary Games

The necessary analogy between evolutionary biology and evolutionary game theory comprises (*fitness* ↔ *payoff*) and (*behaviours* ↔ *strategies*), where *payoff* is a measure of reward (cardinal) or desirability (ordinal) for a given outcome, and *strategy* is a plan of action given the game's state. Evolutionary game theory considers scenarios where agents are not particularly rational, and it is interested in which behaviours are able to persist or are phased out in a population of such agents. In this context, the equivalent of Nash Equilibrium [22] is the so-called *evolutionary stable strategy*. Evolutionary games assume that the number of agents is large, that the agents are anonymous and that the number of populations is finite. Other assumptions are *inertia* and *myopia* which refer to the tendency to switch strategies only occasionally and to base decision-making on local information about one's current neighbourhood, respectively. The shift in perspective from traditional game theory allows to focus on behaviour (of a population) as the distribution of the agents' choices. In other words, one is no longer interested in what strategy did $agent_i$ chose at a given time, but rather how many agents chose strategy s_i over strategy s_j, for example.

There are many types of population games [23], however, this paper focuses on an evolutionary symmetric game with two pure actions.

3.1 Demographic Prisoner's Dilemma

A DPD [24] is a type of evolutionary game where all agents are indistinguishable and they inherit a fixed (immutable) strategy. It is a memoryless game in that each agent, at each *stage game*[1], has no knowledge of the past interactions. It is based on the Prisoner's Dilemma (PD) [25], a two-player symmetric game of imperfect information, shown in Table 1, where c (cooperate) and d (defect) are the two pure actions available to the players, and the tuples in the matrix are the payoffs corresponding to pairwise combinations of these actions, with

---

[1] A single iteration of a repeated game.

$T > R > P > S$. These variables quantify the specific value associated with a given action. Therefore, the most desirable payoff for player X or Y would be attained when defecting while the other player is cooperating. The second best outcome would be to both cooperate, however, the lack of information regarding their respective actions rationally rules out either of these strategies.

Table 1. Prisoner's Dilemma in normal form.

Player Y

		c	d
Player X	c	(R, R)	(S, T)
	d	(T, S)	(P, P)

In fact, for a *one-shot* PD game, it has been shown that the Nash Equilibrium[2] [22] is the pure strategy *dd*. It has also been shown [24] that in a DPD cooperation can emerge and endure, unlike in a repeated PD game with memory, where the dominant strategy would still be to defect. This result not only contradicts some of the axioms of rational decision-making, but also provides an opportunity for speculations at an artistic level (see Sect. 4.1).

Formally, a typical game unfolds as follows: a random number of agents is placed in a discrete spatial grid, and their behaviour is influenced by the surrounding neighborhood. Each agent is born with a fixed strategy, a nominal amount of wealth, age and a vision perimeter. Additionally, it might also have a maximum age limit and a strategy mutation genotype. Each agent looks within its vision perimeter, chooses an available neighbour at random, and plays a PD game with that neighbour. Payoffs add to the agent's wealth which, if exceeding a given threshold, allows the agent to reproduce within its vision perimeter. The child agent inherits the strategy and a portion of the parent's wealth. On the contrary, an agent dies if its wealth falls below zero. These mechanisms, akin to cellular automata, combined with a probabilistic choice of the pairwise matching occurring at each time step of the model, generate a spatial representation of an evolutionary process.

Many variations of the DPD have been proposed and explored, for example by introducing conditional dissociation [26], forced cooperation constraints [27], Markov strategies [28], memory capacity and conditional cooperative strategies [29], and risk attitudes [30]. In this paper, only the vanilla version [24] is considered. To describe the formal analogies between the DPD and the author's AV work, the notion of *conceptual blend* or integration is now explained.

---

[2] The strategy profile that no player has an incentive to change unilaterally.

4   Conceptual Integration

The process of integrating two conceptual spaces onto a third is referred to as conceptual blending [31]. This notion is a useful tool which formalises analogy or metaphor. Intrinsically, this tool has already been used when porting evolutionary biology to evolutionary games in Sect. 3, but it is worth having a more rigorous definition of the integration process. A conceptual blend comprises:

- two *input spaces*: the concepts one wishes to integrate.
- a *generic space*: all the elements in common from the above.
- a *cross-space mapping*: connections of identity, transformation or representation.
- a *blended space*: all the generic structures, as well as structures that might be unattainable or impossible in the inputs.

The projection onto the blended space is selective and it is not unique. How the blended space is arrived at is not a prescriptive procedure, thus "conceptual blending is not a compositional algorithmic process and cannot be modelled as such for even the most rudimentary cases" [32, p. 136]. Figure 1 illustrates a conceptual blend diagram for the AV DPD, while the next section describes the chosen cross-space mapping.

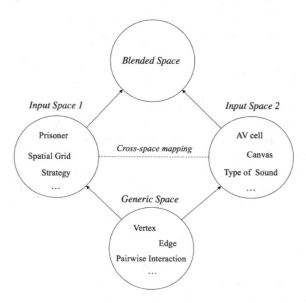

Fig. 1. A possible conceptual blend diagram for the AV DPD

4.1 Audiovisual Demographic Prisoner's Dilemma

Each agent has both visual and audio properties. As for the former, an agent is shown as a cell in a square grid of variable dimensions and it has a different colour depending on its inherited strategy (green for c, red for d). Additionally, its edge roundness is dynamic and changes proportionally to the agent's wealth. On the audio front, each agent is a frequency modulation (FM) [33] unit whose carrier frequency is proportional to a fundamental frequency and the agent's position on the grid. Such fundamental frequency is chosen at the author's discretion before runtime and, similarly to the colour code, it is different depending on the agent's strategy. Defectors are associated with a fundamental frequency below the hearing range, with an added element of randomness, whereas cooperators have a fundamental frequency within the hearing range. Thus, defectors have a 'noisy' sound texture, whereas cooperators are contributing to a harmonic texture which is richer as their number grows. Each active agent's sound component is spatialised using binaural panning.

The correspondence between cooperation and a harmonic sound resulting from the partials contributed by the agents is perhaps simplistic or even cliché in that endorses certain socioeconomic narratives around collaboration. The choice of the colour code is liable of a similar critique. Some of these limitations will become more obvious later on, and will be discussed in Sect. 6. However, these choices are not entirely unreasonable and constituted a sufficient mapping that allowed to explore the population dynamics through auditory and visual display. The several versions of the AV DPD that were implemented are now described.

5 Evolving Prisoners

The first version of the AV DPD does not set an upper bound on the agents' age, nor does it allow for mutation of their offspring's strategy. Emergence of cooperation as an evolutionary stable behaviour could be observed around the 50th iteration.

A second implementation limits the age to an arbitrary value which can be experimented with heuristically. Once and if the whole population dies out, the game is automatically restarted. Age is represented both in the visual domain, as the transparency value (the older the agent, the more transparent it is), and in the audio domain, being mapped to the amplitude of both oscillators for any given active agent.

Yet a third implementation of the game adds a probability that a given child might mutate strategy instead of inheriting the parent's one. In the games shown in Figs. 2 and 3 this probability was set at 0.5, thus rather high according to evolutionary standards, but can be changed arbitrarily.

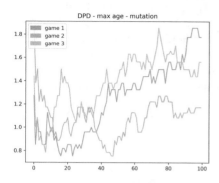

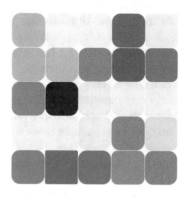

Fig. 2. The ratio between cooperators and defectors in three games (100 iterations shown) of a max age, mutation DPD

Fig. 3. A screen shot of an iteration

6   Aesthetic Pit Stop

In all three versions, cooperation seemed to emerge over defection. Sonically, this behaviour produced a rather minimalist and meditative result, a continuously morphing harmonic texture where different partials were enhanced or suppressed based on the population's behaviour. The superposition of the cooperator's partials over the randomness of the defectors achieved a smooth blend between noise pointillism and a rich, evolving harmonic content.

However, on the visual front, results were far from satisfactory (see Fig. 3). While the mapping applied in Sect. 4.1 proved effective as an aid to visualising and understanding the population dynamics of the DPD, it is important to consider the aesthetic implications of such cross-space mapping. The latter is a paramount factor for many other processes involving the display of data, e.g. sonification. In the context of computational systems which seem to exhibit a level of autonomy, such as population-based systems, it is crucial to review the emerging area of *computational aesthetics*.

6.1   Computational Aesthetics

The theory, practice and applications of aesthetics to computer science in the domains of art, graphics, design and music is normally referred to as computational aesthetics. This notion poses an ontological paradox, suggesting that it might be possible to combine a theory of experience, such as aesthetics, with a model of abstraction, such as computing [34]. The notion of aesthetics in computer science poses interesting and challenging questions, and it has been noted that, regrettably, computational aesthetics has continued to evolve somewhat independently from aesthetics in philosophy of the arts. For more details on the shortcoming of such decoupling the reader is encouraged to refer to [35] and to [36] for computational music aesthetics in particular. Despite these limitations, and even if "the assumed roles of aesthetics as applied to computing are too limiting" [37, p. 12], it is inevitable to consider the implication of these roles when implementing generative systems for the arts. In [38] a fundamental dichotomy between two strands of computational aesthetics is posited.

According to this view, one strand is concerned about a human-based notion of beauty and pleasantness. The other, instead, is aimed at examining the ability of computational systems to evaluate their own process or product. The latter thus investigates machine meta-aesthetics and autonomy in general. In this context, different approaches, methods and aesthetic measures for the automated evaluation of aesthetic value have been proposed. Amongst them, geometric, formulaic, design, biologically, complexity, information theory, psychological and neural network based are the principal. The details of these largely transcend the scope of this paper and the reader is encouraged to refer to [38] for a thorough survey of the developments in this area.

In the specific field of computational aesthetics applied to evolutionary systems, a common approach is to employ automated fitness functions that can be based on the designer's goals (e.g., complexity, geometry and symmetry, etc.), although some have experimented with granting users more agency over these [39]. In the AV DPD there exists an intrinsic fitness function, which corresponds to the payoff matrix (see Table 1). It is worth noting that aesthetic analogies, akin to conceptual blends (see Sect. 4), have also undergone evolutionary treatment, as in the case of [40].

While this is an interesting route to follow, in this paper the author is more interested in the first notion of computational aesthetics, perhaps one closer to an aesthetics of digital arts that to machine meta-aesthetics. The relation between the computational system, its process and product/output, and the impact of this on the author's own notion of aesthetics is therefore discussed in the next section.

7   Evolving Aesthetics

The discrete nature of the grid, combined with the hard-edged colour dichotomy, proved an important factor in the author's perceived violation of his aesthetic expectations. This issue could have been perhaps reduced by augmenting the size of the population. However, when this was attempted, the system's audio processing was often unable to handle the computational load. The AV DPD was implemented using the WebAudio API [41] and anything above 40 agents was compromising the audio processing, because each active agent involves several Audio Nodes (e.g. oscillators, gain, envelopes). While the obvious solution to this problem would have been to implement the AV DPD with a dedicated audio programming software, the author decided to stick to a web browser version for reasons of accessibility (operating system agnostic, no software download required on behalf of potential users) and future integration with personal website.

To arrive at a satisfactory outcome with respect to the author's aesthetic aspirational levels, the cross-space mapping of the AV DPD was gradually modified. In particular, the author sought to match the minimalist and meditative

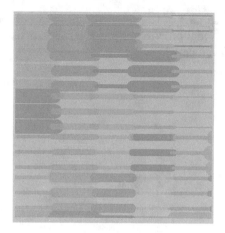

Fig. 4. AV DPD: a cream palette   **Fig. 5.** AV DPD: a grey palette

sonic output of the system to an equivalent visual representation. Thus, the colour code mapping was dramatically changed, in the attempt to minimise the dichotomy of the agent's appearance. Agents (cooperators and defectors) were given similar (but not identical) values and they were drawn with a variable stroke weight and spacing, based on their wealth.

The changes described were minimal, however, the results so attained (see Figs. 4 and 5) were encouraging and were considered *satisficing*[3] [42] by the author.

8   Discussion

The role of the cross-space mapping in conceptual blends between evolutionary game theory and audiovisual works is crucial and often decisive for attaining satisfactory results. Although automated aesthetic measures (e.g., fitness functions) can be used to evaluate the latter, it is arguable that the opinion of domain practitioners or experts is a *sine qua non* for artistic relevance. It has been noted that "the'art' in EA [evolutionary art] seems to be largely taken for granted and is passed with little commentary or analysis from within the field." [35, p. 1] It is thus important to interrogate the notion of computational aesthetics and to understand how this relates to the broader fields of art and philosophy, where universal notions of beauty and pleasantness have long been abandoned. Aesthetics relates to one's affective response to a (creative) process or artefact, and it is thus linked not only to perception and sensory discrimination, but also to considerations of moral, ethical or political nature. Therefore, it is a complex and dialogical experience which can be difficult to boil down to automated measures and procedures. Setting individual aspiration levels and negotiating these

---

[3] Sufficiently satisfactory.

as part of the relational process between the designer/artist and the computational process/output is a fundamental, recursive operation which should not be abdicated.

Fig. 6. Combining the DPD with Conway's Game of Life

9   Future Work

Future directions of the AV DPD include further exploration of the the cross-space mapping but also extensions in the evolutionary scope. For example, agents' phenotype (properties) could be augmented to include more rules of behaviour. Figure 6 shows an example of combining the DPD with the cellular automata known as the Game of Life [43]. Another strategy could be to vary the genotype (e.g., oscillator type, audio parameters, etc.). An example of a variation of the AV DPD could be a simple manipulation of the transparency level assigned to cooperators or defectors, whereby a background image could gradually emerge in correspondence to evolutionary stable strategies. Cooperation could also be evaluated with respect to a given threshold so to unlock specific sonic behaviours of the AV work. Yet other directions for future work could be to explore scale-free networks [44], providing a viable solution is sought with respect to the audio processing requirements, or some other DPD variation mentioned at the end of Sect. 3.1. The AV DPD could also be modified to allow interactivity with (potential) users. For example, the system could provide real-time control over some of the parameters, such as the mutation rate, the colour palette, or the fundamental frequency of the cooperators and/or the defectors.

The possibilities are virtually endless, limited only by the imagination applied to the cross-space mapping on behalf of the designer.

References

1. Lindemann, A., Lindemann, E.: Musical organisms. In: Liapis, A., Romero Cardalda, J.J., Ekárt, A. (eds.) EvoMUSART 2018. LNCS, vol. 10783, pp. 128–144. Springer, Cham (2018). https://doi.org/10.1007/978-3-319-77583-8_9

2. Olseng, O., Gambäck, B.: Co-evolving melodies and harmonization in evolutionary music composition. In: Liapis, A., Romero Cardalda, J.J., Ekárt, A. (eds.) Evo-MUSART 2018. LNCS, vol. 10783, pp. 239–255. Springer, Cham (2018). https://doi.org/10.1007/978-3-319-77583-8_16

3. Loughran, R., O'Neill, M.: Clustering agents for the evolution of autonomous musical fitness. In: Correia, J., Ciesielski, V., Liapis, A. (eds.) EvoMUSART 2017. LNCS, vol. 10198, pp. 160–175. Springer, Cham (2017). https://doi.org/10.1007/978-3-319-55750-2_11

4. Scirea, M., Togelius, J., Eklund, P., Risi, S.: MetaCompose: a compositional evolutionary music composer. In: Johnson, C., Ciesielski, V., Correia, J., Machado, P. (eds.) EvoMUSART 2016. LNCS, vol. 9596, pp. 202–217. Springer, Cham (2016). https://doi.org/10.1007/978-3-319-31008-4_14

5. Eigenfeldt, A., Pasquier, P.: Populations of populations: composing with multiple evolutionary algorithms. In: Machado, P., Romero, J., Carballal, A. (eds.) Evo-MUSART 2012. LNCS, vol. 7247, pp. 72–83. Springer, Heidelberg (2012). https://doi.org/10.1007/978-3-642-29142-5_7

6. Linkola, S., Hantula, O.: On collaborator selection in creative agent societies: an evolutionary art case study. In: Liapis, A., Romero Cardalda, J.J., Ekárt, A. (eds.) EvoMUSART 2018. LNCS, vol. 10783, pp. 206–222. Springer, Cham (2018). https://doi.org/10.1007/978-3-319-77583-8_14

7. McCormack, J.: Niche constructing drawing robots. In: Correia, J., Ciesielski, V., Liapis, A. (eds.) EvoMUSART 2017. LNCS, vol. 10198, pp. 201–216. Springer, Cham (2017). https://doi.org/10.1007/978-3-319-55750-2_14

8. Neumann, A., Alexander, B., Neumann, F.: Evolutionary image transition using random walks. In: Correia, J., Ciesielski, V., Liapis, A. (eds.) EvoMUSART 2017. LNCS, vol. 10198, pp. 230–245. Springer, Cham (2017). https://doi.org/10.1007/978-3-319-55750-2_16

9. Cohen, M.W., Cherchiglia, L., Costa, R.: Evolving mondrian-style artworks. In: Correia, J., Ciesielski, V., Liapis, A. (eds.) EvoMUSART 2017. LNCS, vol. 10198, pp. 338–353. Springer, Cham (2017). https://doi.org/10.1007/978-3-319-55750-2_23

10. Kelly, J., Jacob, C.: evoExplore: multiscale visualization of evolutionary histories in virtual reality. In: Liapis, A., Romero Cardalda, J.J., Ekárt, A. (eds.) EvoMUSART 2018. LNCS, vol. 10783, pp. 112–127. Springer, Cham (2018). https://doi.org/10.1007/978-3-319-77583-8_8

11. Martins, T., Correia, J., Costa, E., Machado, P.: Evotype: towards the evolution of type stencils. In: Liapis, A., Romero Cardalda, J.J., Ekárt, A. (eds.) EvoMUSART 2018. LNCS, vol. 10783, pp. 299–314. Springer, Cham (2018). https://doi.org/10.1007/978-3-319-77583-8_20

12. Lourenço, N., Assunção, F., Maçãs, C., Machado, P.: EvoFashion: customising fashion through evolution. In: Correia, J., Ciesielski, V., Liapis, A. (eds.) EvoMUSART 2017. LNCS, vol. 10198, pp. 176–189. Springer, Cham (2017). https://doi.org/10.1007/978-3-319-55750-2_12

13. Tabatabaei Anaraki, N.A.: Fashion design aid system with application of interactive genetic algorithms. In: Correia, J., Ciesielski, V., Liapis, A. (eds.) EvoMUSART 2017. LNCS, vol. 10198, pp. 289–303. Springer, Cham (2017). https://doi.org/10.1007/978-3-319-55750-2_20

14. Bak, S.H., Rask, N., Risi, S.: Towards adaptive evolutionary architecture. In: Johnson, C., Ciesielski, V., Correia, J., Machado, P. (eds.) EvoMUSART 2016. LNCS, vol. 9596, pp. 47–62. Springer, Cham (2016). https://doi.org/10.1007/978-3-319-31008-4_4

15. Byrne, J., Hemberg, E., Brabazon, A., O'Neill, M.: A local search interface for interactive evolutionary architectural design. In: Machado, P., Romero, J., Carballal, A. (eds.) EvoMUSART 2012. LNCS, vol. 7247, pp. 23–34. Springer, Heidelberg (2012). https://doi.org/10.1007/978-3-642-29142-5_3

16. Xenakis, I.: Formalized Music: Thought and Mathematics in Composition. Harmonologia Series. Pendragon Press, Stuyvesant (1992)

17. Leslie, G.: A game theoretical model for musical interaction. In: Proceedings of the International Computer Music Conference (2008)

18. Canonne, C.: Free improvisation and game theory: an introduction and a few experiments (2008). http://www.satelita.de/downloads/Free_Improvisation_and_Game_Theory.pdf. Accessed 3 May 2018

19. Brackett, J.L.: John Zorn: Tradition and Transgression. Indiana University Press, Bloomington (2008)

20. Kalonaris, S.: Markov networks for free improvisers. In: Proceedings of the 42nd International Computer Music Conference, Utrecht, The Netherlands, pp. 182–186 (2016)

21. Kalonaris, S.: Adaptive specialisation and music games on networks. In: Proceedings of the 13th International Symposium on Computer Music Multidisciplinary Research, Matosinhos, Portugal, pp. 420–429 (2017)

22. Nash, J.F.: Equilibrium points in n-person games. Proc. Natl. Acad. Sci. **36**(1), 48–49 (1950)

23. Sandholm, W.H.: Population games and deterministic evolutionary dynamics. In: Handbook of Game Theory with Economic Applications, vol. 4, no. 1, pp. 703–778 (2015)

24. Epstein, J.M.: Zones of cooperation in demographic prisoner's dilemma. Complexity **4**(2), 36–48 (1998)

25. Poundstone, W.: Prisoner's Dilemma: John Von Neumann, game theory and the puzzle of the bomb. Phys. Today **45**, 73 (1992)

26. Izquierdo, L.R., Izquierdo, S.S., Vega-Redondo, F.: Leave and let leave: a sufficient condition to explain the evolutionary emergence of cooperation. J. Econ. Dynam. Control **46**, 91–113 (2014)

27. Zhang, H., Gao, M., Wang, W., Liu, Z.: Evolutionary prisoner's dilemma game on graphs and social networks with external constraint. J. Theor. Biol. **358**, 122–131 (2014)

28. Stahl, D.O.: Cooperation in the sporadically repeated prisoners' dilemma via reputation mechanisms. J. Evol. Econ. **21**(4), 687–702 (2011)

29. Moreira, J., et al.: Individual memory and the emergence of cooperation. Anim. Behav. **85**(1), 233–239 (2013)

30. Zeng, W., Li, M., Chen, F.: Cooperation in the evolutionary iterated prisoner's dilemma game with risk attitude adaptation. Appl. Soft Comput. **44**, 238–254 (2016)

31. Fauconnier, G.: Conceptual blending. In: Smelser, N.J., Baltes, P.B. (eds.) Int. Encycl. Soc. Behav. Sci., pp. 2495–2498. Pergamon, Oxford (2001)

32. Fauconnier, G., Turner, M.: Conceptual integration networks. Cogn. Sci. **22**(2), 133–187 (1998)

33. Chowning, J.M.: The synthesis of complex audio spectra by means of frequency modulation. J. Audio Eng. Soc. **21**(7), 526–534 (1973)

34. Fazi, M.B.: Incomputable aesthetics: open axioms of contingency. Comput. Cult. **5** (2016). http://computationalculture.net/incomputable-aesthetics-open-axioms-of-contingency/. Accessed 3 May 2018

35. McCormack, J.: Aesthetics, art, evolution. In: Machado, P., McDermott, J., Carballal, A. (eds.) EvoMUSART 2013. LNCS, vol. 7834, pp. 1–12. Springer, Heidelberg (2013). https://doi.org/10.1007/978-3-642-36955-1_1

36. Kalonaris, S., Jordanous, A.: Computational music aesthetics: a survey and some thoughts. In: Proceedings of the 3rd Conference on Computer Simulation of Musical Creativity, Dublin, Ireland (2018)

37. Fishwick, P.: An introduction to aesthetic computing. In: Aesthetic Computing, pp. 3–27. The MIT Press (2006)

38. Galanter, P.: Computational aesthetic evaluation: past and future. In: McCormack, J., d'Inverno, M. (eds.) Computers and Creativity, pp. 255–293. Springer, Heidelberg (2012). https://doi.org/10.1007/978-3-642-31727-9_10

39. Machado, P., Martins, T., Amaro, H., Abreu, P.H.: An interface for fitness function design. In: Romero, J., McDermott, J., Correia, J. (eds.) EvoMUSART 2014. LNCS, vol. 8601, pp. 13–25. Springer, Heidelberg (2014). https://doi.org/10.1007/978-3-662-44335-4_2

40. Breen, A., O'Riordan, C., Sheahan, J.: Evolved aesthetic analogies to improve artistic experience. In: Correia, J., Ciesielski, V., Liapis, A. (eds.) EvoMUSART 2017. LNCS, vol. 10198, pp. 65–80. Springer, Cham (2017). https://doi.org/10.1007/978-3-319-55750-2_5

41. Web Audio API. https://developer.mozilla.org/en-US/docs/Web/API/Web_Audio_API. Accessed 5 Nov 2018

42. Simon, H.: Rational choice and the structure of the environment. Psychol. Rev. **63**(2), 129–138 (1956)

43. Gardner, M.: Mathematical games: the fantastic combinations of John Conway's new solitaire game "life". Sci. Am. **223**, 120–123 (1970)

44. Du, W.B., Zheng, H.R., Hu, M.B.: Evolutionary prisoner's dilemma game on weighted scale-free networks. Physica A Stat. Mech. Appl. **387**(14), 3796–3800 (2008)

Emojinating: Evolving Emoji Blends

João M. Cunha[(✉)] ⓘ, Nuno Lourenço ⓘ, João Correia ⓘ, Pedro Martins ⓘ,
and Penousal Machado ⓘ

CISUC, Department of Informatics Engineering, University of Coimbra,
Coimbra, Portugal
{jmacunha,naml,jncor,pjmm,machado}@dei.uc.pt

Abstract. Graphic designers visually represent concepts in several of
their daily tasks, such as in icon design. Computational systems can be
of help in such tasks by stimulating creativity. However, current computational approaches to concept visual representation lack in effectiveness
in promoting the exploration of the space of possible solutions. In this
paper, we present an evolutionary approach that combines a standard
Evolutionary Algorithm with a method inspired by Estimation of Distribution Algorithms to evolve emoji blends to represent user-introduced
concepts. The quality of the developed approach is assessed using two
separate user-studies. In comparison to previous approaches, our evolutionary system is able to better explore the search space, obtaining
solutions of higher quality in terms of concept representativeness.

Keywords: Evolutionary Algorithm · Emoji ·
Interactive Evolutionary Computation · Visual blending

1 Introduction

In the domain of visual representations, computers have been made to draw on
their own (e.g. [1]), used as creativity support tools for drawing (e.g. [2]) and even
been given the role of colleague in co-creative systems (e.g. [3]). These examples,
however, are more related to Art and shift away from the Design domain, in
which a specific problem is addressed – e.g. how to design an icon to represent
a given concept.

The difficulty behind developing computational approaches to solve Design
problems is that, in most cases, they greatly depend on human perception. For
this reason, they can be seen as open-ended as there is no optimal solution since
they hinge on the user preferences. Thus, assessing quality is a complex problem
on its own. One possible way to tackle this problem is to develop a system that
allows the user to choose which solutions are adequate. One of such approaches
in the Evolutionary Computation domain is usually referred to as Interactive
Evolutionary Computation (IEC). IEC has been seen as suitable for such open-
ended design problems [4], since it is capable of accumulating user preferences
and, at the same time, stimulating creativity.

ⓒ Springer Nature Switzerland AG 2019
A. Ekárt et al. (Eds.): EvoMUSART 2019, LNCS 11453, pp. 110–126, 2019.
https://doi.org/10.1007/978-3-030-16667-0_8

Regarding the visual representation of concepts, a multi-purpose system has great potential to be used as an ideation-aiding tool for brainstorming activities, presenting the user with representations for concepts. In [5], we have presented such a system. It uses a dataset of input visual representations (emoji), which are combined using a visual blending process to represent new concepts. Despite being able to achieve a higher conceptual coverage than the one from emoji system [6], the implemented system does not employ an effective strategy for exploring the search space – it only considers the best semantically matched emoji for blend generation. This approach ignores most of the search space and does not guarantee that the solutions are the most satisfactory for the user – one of the shortcomings identified in [5].

In this paper, we tackle the aforementioned issues by proposing a IEC framework that combines a standard Evolutionary Algorithm with a method inspired by Estimation of Distribution Algorithms to evolve visual representations of user-introduced concepts. We conduct two user-studies to compare our results with the ones reported in [5,6].

2   Related Work

In this paper, we present a visual blending system for concept representation that uses an interactive evolutionary approach. As such, our work addresses two different topics: Visual Representation of concepts and Interactive Evolutionary Computation. In this section, we will present related work for both topics.

2.1   Visual Representation of Concepts

There are different approaches to the visual representation of concepts. One of the approaches consists in gathering a set of individual graphic elements (either pictures or icons), which work as a translation when put side by side – e.g. translating plot verbs into sequences of emoji [7] or the *Emojisaurus* platform[1].

Other approaches are based on input visual representations, which are used to produce new ones. The co-creative system Drawing Apprentice [3], for example, uses convolutional neural networks to perform real-time object recognition on the user sketch and responds with drawings of related objects. Ha and Eck [8] use sketches drawn on *Quick, Draw![2]* to train a recurrent neural network capable of generalising concepts in order to draw them. The system is able to interpolate between several concepts (e.g. a pig, a rabbit, a crab and a face), which can be used for representing new concepts through visual blending – e.g. [9].

Visual blending consists in merging two or more visual representations to produce new ones. Approaches to visual blending can be divided into two groups based on the type of rendering used: photorealistic or non-photorealistic. One example from the photorealistic group is the system Vismantic [10], which uses

---

[1] emojisaurus.com, retrieved 2019.
[2] quickdraw.withgoogle.com, retrieved 2019.

three binary image operations (juxtaposition, replacement and fusion) to produce visual compositions for specific meanings (e.g. fusion of an image of an electric light bulb with an image of green leaves to represent *Electricity is green*). Another photorealistic example is the generation of new faces from existing ones, by combining face parts [11]. Non-photorealistic examples are the generation of visual representations for *boat-house* [12] or for the blends between the concepts *pig/angel/cactus* [13], which combine input visual representations to represent the new concepts. While these explorations only address a reduced number of concepts, the system presented in [5] – upon which this paper builds – works on a bigger scale by combining Semantic Network exploration with visual blending to automatically represent user-introduced concepts using emoji.

2.2   Interactive Evolutionary Computation

Evolutionary Algorithms (EAs) are computational models inspired by the Theory of Natural Selection. They are normally used in problems in which it is possible to assess the quality of solutions based on a specific goal. However, for problems in which quality is highly subjective and dependent on human perception, approaches that involve the user in the evolutionary process are seen as better suited [4]. Such approaches are often referred to as Interactive Evolutionary Computation (IEC) and are characterised by having a user-centred evaluation process. IEC has been used in different domains such as: Fashion, to produce shoe designs according to user taste [14]; Poster Design, to evolve typographic posters [15]; or even Information Visualisation, to explore the aesthetic domain [16]. In terms of symbol generation and visual representation of concepts, IEC has also been seen as a possible approach to solve problems.

Dorris et al. [17] used IEC to evolve anthropomorphic symbols that represented different emotions (e.g. anger, joy, etc.). The genotype of each individual was a vector of nine real-valued numbers that corresponded to the angles of the nine limbs (e.g. torso, left shoulder, right elbow, etc.). Dozier et al. [18] focused on emoticon design using an interactive distributed evolutionary algorithm – multiple processors working in parallel. It allowed several participants to interact in simultaneously, evolving solutions that are the result of their judgements. The emoticons were represented as a vector of 11 integer variables, which corresponded to the y-coordinates of 11 points (e.g. the first three codified the left eyebrow). Piper [19] also used a distributed approach, proposing an interactive genetic algorithm technique for designing safety warning symbols (e.g. *Hot Exhaust*). It used previously drawn symbol components as input which were then combined to produce new symbols. According to Piper [19], this distributed approach allowed the replacement of the usual focus group in symbol design process with a group of participants interacting using computers in a network. Hiroyasu et al. [20] proposed an interactive genetic algorithm that uses a crossover method based on probabilistic model-building for symbol evolution according to user preference. Each individual (symbol) was a combination of a color (HSB system) and a shape (from a set of eight different shapes). Estimation of distribution algorithms (EDAs) are based on the idea that statistical information about the search space can be extracted and used to modify the probability model, reducing the search space and leading faster to good solutions [21,22]. Our approach takes inspiration from EDA methods to provide a way to quickly and efficiently search for solutions that match the user preferences.

3 Background

Emoji are pictorial symbols that are well integrated in the written language, which is observed in the growing number of emoji-related tools and features – e.g. search-by-emoji supported by Google[3], and the Emoji Replacement and Prediction features implemented in iOS 10[4].

Due to their large conceptual coverage (currently there are 2823 emoji in the released Emoji 11.0) and associated semantic knowledge, they are suitable to be used in computational approaches to the visual representation of concepts. In [5], we presented a system that uses a combinatorial approach to represent concepts using emojis as visual representations. It integrates data from three online open resources: Twitter's Twemoji 2.3: a dataset of emoji fully scalable vector graphics, in which each emoji is composed of several layers; EmojiNet, a machine readable sense inventory for emoji built through the aggregation of emoji explanations from multiple sources [23]; and ConceptNet: a semantic network originated from the project Open Mind Common Sense [24].

The system has three components: (i) the *Concept Extender* (CE) searches related concepts to a given concept using ConceptNet; (ii) the *Emoji Searcher* (ES) searches for existing emoji semantically related to a given word; and (iii) the *Emoji Blender* (EB) produces new emoji using visual blending by adding/replacing layers. For a more detailed description, we refer the reader to [5].

In [6], we assessed the system's performance using 1509 nouns from the New General Service List (NGSL) [25] and reported that the system was able to produce emoji for 75% of the list. Despite considering these results as good, the tested system only uses the best semantically matched emoji for each concept, not exploring the full search space. With this in mind, we propose an evolutionary approach to explore the search space and find visual representations of concepts that match user preference.

4 Our Approach

In the context of this project, we use the *Emoji Searcher* (ES) and the *Concept Extender* (CE) components presented in [5], and we introduce an novel approach to explore the search space using IEC methodologies. As such, an evolutionary

---

[3] forbes.com/sites/jaysondemers/2017/06/01/could-emoji-searches-and-emoji-seo-become-a-trend/, retrieved 2018.

[4] macrumors.com/how-to/ios-10-messages-emoji/, retrieved 2018.

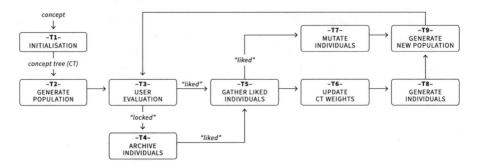

Fig. 1. Evolutionary framework diagram, showing tasks (T1-9) and objects, e.g. *concept tree* (CT). The user starts by introducing a concept, which will be used to generate a random set of solutions (T1 and T2). Then, the user evaluates the individuals by selecting the ones that fit his/her preferences (T3), referred to as "liked individuals". The user can also select the individuals to be stored in an archive (T4), referred to as "locked individuals". After the evaluation, the fittest individuals ("liked") are gathered from the population and from the archive (T5). The gathered individuals are then used to produce offspring through mutation (T7), and to update the weights of the Concept Tree (T6) – a graph-structured object with which new individuals are generated (T8). This process can be repeated indefinitely until the user is satisfied.

system was implemented. The *Emoji Blender* (EB) component was modified in order to work together with the evolutionary engine, in which the generated blends are the phenotype of individuals.

The approach presented in this paper has a two-level evolution: on a macro level, it uses a method that takes inspiration from EDAs to direct the search to areas that match the user preference; on a micro and more specific level, it uses a standard EA to focus the evolution on certain individuals. The approach is schematically represented in Fig. 1.

4.1   Concept Tree and General Evolution

The ES and CE components are used together to produce a graph-like structured object from the user introduced concept (T1 in Fig. 1). This object – which we will refer to as Concept Tree (CT) – stores the conceptual and emoji data produced from the analysis of the concept (see Fig. 2). It has two different levels: concept level, in which related concepts are connected (e.g. the concept *god* is connected with the related concepts *judge people*, *judge men*, *justify hate* and *quiet storm*); and the emoji level, which stores the sets of emoji retrieved for each concept (e.g. *judge* has a set of emoji and *men* has another, see Fig. 2).

The complexity of the CT object depends on the type of concept introduced by the user. If it is a single-word concept (e.g., *god*), related double-word concepts are searched and then corresponding emoji are retrieved for each of the words. If the user introduces a double-word concept, no related concept is required and the system directly retrieves emoji.

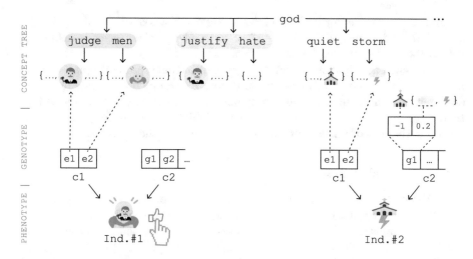

Fig. 2. Individual Representation and weight update system. The figure shows the two chromosomes (*c1* and *c2*) of two individuals' genotypes. It also shows gene 1 (*g1*) of *c2* from individual #2 in detail. Regarding the weight system, the individual #1 is being "liked", which directly increases the weights of the concepts/emoji marked in green and indirectly increases the ones of concepts/emoji marked in grey.

Taking inspiration from EDA methods, a weight value is assigned to every concept in the set of related concepts (in case they exist) and to each emoji. These weights are also stored in the CT object. When we generate new individuals (T2 and T8), the weights are used to select both the concept and the two emoji for the new individual – the higher the weight, the greater chances it has of being selected. Initially the weights are all set to 1 and are updated in each generation according to user preferences (T6 in Fig. 1).

4.2   Representation

The emoji from the Twitter's Twemoji dataset are composed of layers – e.g. the *storm* emoji in Fig. 2 has two layers, a *cloud* and a *lightning*. As already mentioned, each visual blend is the phenotype of an individual (see Fig. 2). The individuals are encoded using a two chromosome genotype, which codify the combination between two emoji. The first chromosome (*c1* in Fig. 2) stores the two emoji (*e1* and *e2*) used in the blend. The second chromosome (*c2* in Fig. 2) is responsible for defining the exchanges of parts that occur in the blend. This is codified by having each exchange stored as a gene (e.g. *g1*). Each gene corresponds to a set of two values: the first defines the part from *e1* that will be used as replacement (−1 in Fig. 2) and the second corresponds to the layer that will be replaced in *e2* (0.2 in Fig. 2). As the number of layers is not the same among emoji, we use numbers in the [0, 1] interval, which correspond to the position of the layer in the layer array. The value −1 can also be used, when the whole emoji is to be used instead of a layer (e.g. when a juxtaposition blend occurs). For example, for individual #2 in Fig. 2 the whole *church* emoji is used as replacement (encoded by the "−1" of *g1*) and the *cloud* layer is defined as replaceable part (encoded by the "0.2" of *g1*).

In the current implementation, two types of blends are used: replacement (a part of *e1* is replaced by *e2*) and juxtaposition (*e1* and *e2* are put side by side or one over the other). In both cases, only one exchange is encoded per individual. In the future, we plan to implement a third blend type (fusion), which uses several exchanges of parts – already supported by the chosen representation.

4.3   User Evaluation

In each generation, a population of 20 individuals is presented to the user, who is able to perform two different actions that affect the evolutionary process: mark individuals as "liked", which increases their fitness; and store them in the archive.

When an individual is "liked" (e.g. Ind.#1 in Fig. 2), the weights of the CT are updated (T6). It directly increases the weight of the related concept behind the individual and of the used emoji in the sets belonging to the concept (marked in green in Fig. 2). A process of indirect weight assignment is also used as the system searches for the used emoji in other concept's sets and also increases the weight of the emoji and corresponding concept (marked in grey in Fig. 2). This fosters related concepts that also use the same emoji and allows the system to find solutions that might also be of interest to the user. The weight increment is calculated based on the sum of weights of a set and it varies according to the type – a direct increment is 5% of the weight sum and an indirect is 2%. In order to make the evolutionary system work, the user does not need to classify every single candidate solution but only select the ones considered interesting.

4.4   Weight Equalisation

A method of weight equalisation was implemented, which means that, as the evolutionary process progresses, there is a tendency towards an equal distribution of weights. The weights between concepts and between emoji inside sets will eventually converge to the same value, if not stimulated. This allows the system to achieve diversity even in late generations, avoiding unwanted convergence.

First, the average of weights inside a set is calculated. The weights that are above average are updated according to the following function (EQ_RATE = $\frac{1}{3}$):

$$new\_weight = (current\_weight - average\_weight) \times \text{EQ\_RATE} \qquad (1)$$

The method of weight equalisation is particularly useful when used together with an archive, which allows the user to increase population diversity and explore other areas of the search space, without losing individuals deemed as good. Despite being different from classic EAs (in which convergence is a goal), this approach fits the problem, as the goal is to help the user find the highest number of interesting, yet considerably different, solutions.

4.5   Archive

An archive is often used to avoid the loss of population diversity by storing good individuals that can later be reintroduced in the population – e.g. [14, 26]. Another possible use is to store the fittest individuals in order to use them to automatically guide the search towards unseen solutions – e.g. [27].

In our case, diversification of the population is achieved with weight equalisation (as already described). Our archive works as a storage of individuals and has two main functionalities: (i) to save individuals and to avoid losing them in the evolutionary process; (ii) allowing the user to activate a permanent "liked" status that leads to a constant fostering of individuals. This option helps in situations that the user has found an individual with an interesting trait (e.g. the use of a specific emoji) and wants to constantly stimulate it without having to manually do it on each generation. As explained before, selecting an individual as "liked" not only fosters its specific evolution but also has an effect on general evolution – changing CT weights and consequently affecting the generation of new individuals.

Moreover, storing the individual in the archive is the only way of guaranteeing that it is not lost when moving to the next generation. It allows the user to store good solutions and focus on other possibilities, while being able, at any time, to further evolve the stored individual, by activating the "liked" option. Combined with the weight equalisation, this makes it possible for the system to increase its diversity and, at the same time, avoid the loss of good individuals. This strategy allows the user to continuously change its exploration goal and try to find new promising areas in the search space.

4.6   Mutation

In addition to being used to update the CT weights (T6), user-evaluated individuals (the "liked" ones) are also employed in the production of offspring in each generation (T7). These are gathered from both the current population and the archive, being the individuals marked with "like". From each "liked" individual, a set of 4 new individuals are produced (e.g. in Fig. 3, the 4 "bread-rhinos" in the population were generated by mutating the "liked" one in the archive). The parent individual goes through a mutation process, in which three types of mutation may occur: (i) emoji mutation (20% probability of occurring) – the emoji used as replacement is changed; (ii) layer mutation (80% probability of occurring per gene) – the replaced layer is changed (e.g. all "bread-rhinos" in the population except the first); and (iii) blend type mutation (5% probability) – this mutation changes the type of blend to juxtaposition, in which the emoji are used together and no replacement occurs (e.g. the first "bread-rhino" in the population). If a blend type mutation happens, no layer mutation occurs. The values presented were empirically obtained through experimentation and adjustments.

The use of the layer and emoji mutation types covers two situations: (i) adequate emoji are being used but the layer is not the correct; (ii) the exchange of layers is considered good but using different emoji may lead to a better solution.

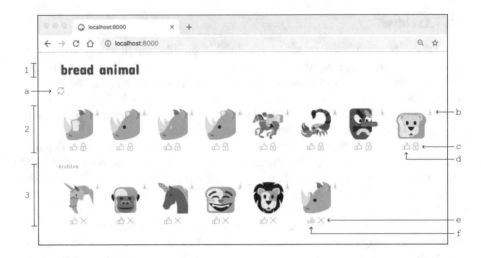

Fig. 3. The interface is divided into 3 areas: *search area* (1), *population area* (2) and *archive area* (3). There are 5 different button types that allow the user to interact with the system: *next generation* (a), *download* (b), *lock* (c), *like* (d) and *remove from archive* (e). A "liked" individual has an activated *like* button (f). The number of individuals in the population was intentionally reduced to increase the legibility of the figure.

4.7   Offspring

The offspring produced from parent individuals (T7) are added to a pool, from which they are afterwards randomly selected for the next generation. The number of individuals in the population is constant (20). As such, there is a maximum percentage of the new population (30%) that is used for individuals generated from parents through mutation. The remaining percentage corresponds to new individuals generated from the CT (T8). When generating individuals from scratch using CT, the probability of juxtaposition is set to 20% and of replacement to the remaining 80% – replacement can lead to many more different solutions than juxtaposition and, as such, it should occur more frequently.

4.8   Interface

The IEC system was implemented as a web-based application, which allows user interaction (see Fig. 3). The interface has three areas: the *search area*, the *population area* and the *archive area* (1–3 in Fig. 3). The *search area* is where the user introduces the concept (e.g. *bread animal* in Fig. 3).

The *population area* presents the current population, showing the visual representation of the blends. Each individual has buttons: the "like", which is used to evaluate the individual (*d*); the "lock", which stores the individual in the archive (*c*); and one to download the visual representation of the individual (*b*).

Individuals in the *archive area* also have a "like" button, which is used to active/deactivate the evaluation of the individual (the choice is maintained between generations), and a button to remove it from the archive (*e* in Fig. 3).

Fig. 4. Blends used in user-survey #1 for the concepts (1–9) *frozen flower, secrets in the future, silent snake, storm of the teacher, the darkest rose, the flame of swords, the laughing blade, the sexy moon* and *the sharp silk*. Blends in the top row are from [5] and the ones in the bottom row are obtained with our system.

Table 1. User-study #1 results expressed in percentage for each concept.

#	Concept	Answers (%)			
		Our image	"Equally good"	Image by [5]	"None"
1	*frozen flower*	54.8	12.9	16.1	16.1
2	*secrets in the future*	9.7	0	58.1	32.3
3	*silent snake*	12.9	22.6	61.3	3.2
4	*storm of the teacher*	22.6	9.7	58.1	9.7
5	*the darkest rose*	9.7	16.1	16.1	58.1
6	*the flame of swords*	0	6.5	90.3	3.2
7	*the laughing blade*	45.2	12.9	16.1	25.8
8	*the sexy moon*	19.4	0	64.5	16.1
9	*the sharp silk*	32.3	3.2	3.2	61.3

5 Results and Discussion

In this section, we present and discuss the experimental results obtained from two user-studies. User-study #1 compares the results from our approach with the ones presented in [5], for a set of concepts. User-study #2 assesses the efficiency of the system in the production of visual representations for single-word concepts and compares the results with the ones described in [6]. In the last part of this section, we make a more general analysis of the approach.

5.1 User-Study #1: Comparing Images from Two Approaches

The first study was used to assess if our approach could lead to better solutions than a non-evolutionary deterministic version of the system, as presented in [5].

In [5], the system was used by 22 participants to generate visual representations for a set of ten concepts and the best solutions were collected. We used our system to produce solutions for the same concepts (see Table 1) and conducted a survey with 31 participants to assess if our system produced better solutions.

We used a multiple choice survey, with one question for each concept, following the model *Which of the following images represents better: [insert concept]?* Each question had four randomly ordered answer options: image produced by our system, image from [5], "equally good" and "none". In order to produce the survey, two people used the system together to select a good representation for each concept (see Fig. 4). Despite the risk of introducing bias towards our preferences, this was a necessary step to reduce the number of options presented to the user. One of the concepts (*serpent of the year*) was not used because the users could not find any good solution different from the one presented in [5].

The survey was conducted with 31 participants, with age between 18–32. The results are shown in Table 1. We can see that for two of the concepts (*frozen flower* and *the laughing blade*) our image was selected as better by the majority of the participants and for *the darkest rose*, 25.8% selected it as better or "equally good". Moreover, for *the sharp silk*, despite the majority of the participants selecting the option "none" (consistent with previous results [5]), our image still had better results than the image presented in [5], which was only selected by 1 participant. All in all, our approach was competitive in 4 out of the 10 concepts.

5.2 User-Study #2: Testing with Concepts from NGSL

In user-study #2, we compare the results from our approach with the ones described in [6], in which a set of 1509 nouns from the New General Service List (NGSL) [25] was used to assess the quality of the system presented in [5]. In the study described in [6], the system allowed the user to introduce a concept and presented an *existing emoji*, *related emoji* (gathered from semantic search to related concepts) and *blends*. It was assessed if each noun was represented by its (i) *existing emoji* and by its (ii) *related emoji* or *blends*. Based on the results presented in [6], we divided the noun list into four groups:

- *group 0:* the system was not able to produce blends. This group was excluded as it could not be used due to the lack of blends to evolve;
- *group 1:* system produces blends but neither the related emoji/blends nor *existing emoji* were reported to represent the concept;
- *group 2:* system produces blends and only the related emoji/blends were reported to represent the concept;
- *group 3:* system produces blends and the existing emoji were reported to represent the concept (the related emoji/blends may also represent).

Moreover, we crossed the list with a dataset of concreteness ratings [28], obtaining a value of concreteness for each noun – from 1 (abstract, language based) to 5 (concrete, experience based). Despite not being a goal in this paper, we divided each noun group in three subgroups to assess if there is any relation between concreteness and representation easiness: (A) low concreteness, (B) medium concreteness and (C) high concreteness.

We conducted a survey with eight participants in which each participant used the system to generate visual representations for 9 randomly selected concepts (one from each subgroup). As the goal for this survey was to achieve maximum coverage of each subgroup, we decided to avoid noun repetition. Despite this, in low concreteness subgroups only few nouns existed – subgroup 1A had 4 nouns, 2A had 3 and 3A had 5 – which led to the repetition of nouns among participants for those subgroups. The participants used the system to evolve visual representations for the nouns, conducting only one run per noun and having a limit of 30 generations. They were asked to find individuals that represented the introduced noun and were allowed to stop the run before reaching 30 generations if they were already satisfied or if the system was not being able to further improve. For each noun, they were also requested to evaluate how well it was represented by the system, from 1 (very bad) to 5 (very good), and exported the solutions that they considered the best among the ones that represented the noun (see Fig. 5).

Table 2. User-study #2 results for *quality, number of solutions, number of generations* and three combinations of quality (Q)/exported (E)/generations (G) that correspond to "early quit without results", "early quit with poor results" and "early satisfaction" (expressed in number of nouns and divided by noun group).

	Quality			# Exported			# Generations			$E = 0$ & $G < 20$	$Q \leq 3$ & $E > 0$ & $G < 20$	$Q \geq 4$ & $E > 0$ & $G < 20$
	1	>1 & ≤3	≥4	0	1	>1	<15	≥15 & <30	30			
group 1	8	6		10	6	8	10	10	7	7	3	4	6
group 2	9	5		10	6	11	7	8	10	6	4	2	7
group 3	6	3		15	4	12	8	13	10	1	4	3	10

The results in Table 2 show that, in terms of quality, our system is able to produce solutions with quality equal or above "good" for almost half of the concepts in group 1 and 2 (10 out of 24), and for the majority of concepts in group 3 (15 out of 24). This is particularly important in group 1, for which the previous study [6] was not able to find any satisfactory solution. Moreover, the participants were able to find more than one concept-representative solution in 34% of the runs (25 out of 72), e.g. *invitation* in Fig. 5.

We were able to compare the individuals selected as the best by each participant for each concept with the solutions obtained by the system from [6] for the same concepts. In 38 out of 72 runs, the solution considered as the best was not produced by the approach from [6]. In addition, in 30 cases out of the 38 our solution was considered better than any of the solutions obtained with the system from [6] and in 5 was considered equally good. This shows that the evolutionary approach has clear advantages in comparison to the approach presented in [5,6].

Concerning the number of generations, in 80% of the runs (58 out of 72) the participants stopped before reaching the generation limit, which can be indicative of two things: the system could not create blends that represented the concept or the user was already satisfied. To further analyse this matter, we used three combinations of quality/exported/generations that correspond to "early quit without results", "early quit with poor results" and "early satisfaction" (see Table 2). From the results we can see that in 11 runs, the participant stopped without any exported solution before reaching 20 generations, which indicates that the system was not being successful. In addition, the column corresponding to "early quit with poor results" shows that in 9 runs the participant considered that the system would not get any better. On the other hand, in 30% of the runs (23 out of 72) the participant was satisfied before reaching the 20th generation, which means that the system was able to quickly evolve solutions that pleased the user.

faith   log   income   kind   obligation   invitation   anything   aircraft

Fig. 5. Examples of blends selected by the participants as good solutions

One of the problems in IEC approaches is the weariness of the user [14]. In the end of the survey, the participants evaluated the weariness degree of the task from 1 (very low) to 5 (very high) and 50% of participants rated it as very low in weariness and the other 50% as low. We also asked the participants to evaluate the surprise degree from 1 (very low) to 5 (very high) – 25% rated it as 3 and 75% as 4. This shows that the system is able to generate solutions that are unexpected. However, as results are highly dependent on the concept, in a proper evaluation the surprise degree should be assessed for each concept.

When analysing the results, we could not observe any obvious relation between concept concreteness and easiness of representation. Our initial expectation was that concrete concepts would be easier to represent. The fact that we could not observe such correlation may indicate that using emoji blending to represent concrete concepts (e.g. *brain*) might not be the best approach. Moreover, some of the participants commented that they were trying to isolate an emoji, which is observed in some of the selected solutions – they tend to mostly show only one of the emoji (see blends for the concepts *anything* and *aircraft* in Fig. 5). However, further research is required on this subject as our remarks are only speculative and not statistically proven.

Another subject concerns the methods used in blend production. For single word concepts, the system gathers related double-word concepts to use in the blending process. The emoji belonging to each of the related concepts are not transferable to other concepts unless they are also in the emoji list of the concept, i.e. two individuals produced from different related concepts, one produced using *emoji* A and B and the other produced using *emoji* C and D, may never lead to the generation of an individual from *emoji* A and D. This is the reason behind some of users complaining that they were not being successful in "combining" emoji from two individuals. This is not very intuitive when using an IEC approach and should be addressed in the future.

5.3   General Analysis

The main goal behind a system for the visual representation of concepts is being able to produce at least one good solution. Our evolutionary system allows the user to explore different and unrelated areas of interest in the search space, often leading to several distinct solutions of good quality for the same concept.

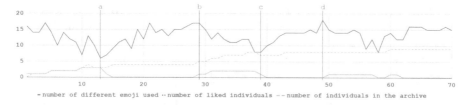

Fig. 6. Metrics progression along the generations of a run for the concept *cell fish* (best viewed in colour). A video of the run can be seen at https://rebrand.ly/evomusart19.

To give an example of how the system reacts to user interaction, we show the progression of several metrics during one run (Fig. 6). It can be observed that the number of different emoji tends to decrease when solutions are marked as "liked", which shows that the population evolves towards similar solutions (e.g. from *b* to *c* in Fig. 6). The opposite is also verified: when no individual is "liked", the variation of the population tends to increase (e.g. from *a* to *b* and from *c* to *d* in Fig. 6). The increase in number of individuals in the archive highlights its usefulness for search space exploration and reflects the capability of the system to evolve solutions that match user preferences.

In the cases in which the system was reported to not being able to generate anything that represented the concept, the reason was related to the gathering of semantic knowledge and had nothing to do with the evolutionary engine (the main focus of this paper). In general, the efficiency of the system is highly dependent on the existing semantic knowledge, emoji found and user perception.

6   Conclusion and Future Work

The visual representation of concepts plays a central role in the work of graphic designers, being crucial in tasks such as icon design. Computational systems can be used to aid the designer in ideation processes by stimulating creativity. In this paper, we proposed an evolutionary approach for the generation of visual representations of concepts that combines a standard Evolutionary Algorithm with a method inspired by EDAs. This approach allows the system to perform both a general evolution to direct the search to areas that match user preference and a focused evolution based on user-selected individuals. In order to do this, we used an archive to store individuals and selectively enable/disable their evolution.

We compared our approach with existing ones, by conducting two user-studies. The results show that our approach allows the exploration of more of the search space and is able to present the user with better solutions. Future enhancements to the proposed approach include: (i) taking into account the semantic value attributed to related concepts and to emoji by the ES and CE components in the initialisation of weights, which may increase the fitness of the population in the first generation; (ii) considering fusion blend type and (iii) a mutation operator for layer transformations (scale, translation and rotation); (iv) using other aspects as ground for blending (e.g. conceptual information related to image schemas [29]); and (v) implementing automatic fitness and (vi) long-term learning (i.e. considering knowledge from past runs).

Demo video: https://rebrand.ly/evomusart19

Acknowledgments. João M. Cunha is partially funded by Fundação para a Ciência e Tecnologia (FCT), Portugal, under the grant SFRH/BD/120905/2016; and is based upon work from COST Action CA15140: ImAppNIO, supported by COST (European Cooperation in Science and Technology): www.cost.eu. This work includes data from ConceptNet 5, which was compiled by the Commonsense Computing Initiative and is freely available under the Creative Commons Attribution-ShareAlike license (CC BY SA 4.0) from conceptnet.io.

References

1. McCorduck, P.: Aaron's code: meta-art, artificial intelligence, and the work of Harold Cohen. Macmillan (1991)
2. Lee, Y.J., Zitnick, C.L., Cohen, M.F.: Shadowdraw: real-time user guidance for freehand drawing. ACM Trans. Graph. (TOG) **30**, 27 (2011)
3. Davis, N., Hsiao, C.P., Singh, K.Y., Magerko, B.: Co-creative drawing agent with object recognition. In: AIIDE 2016 (2016)
4. Parmee, I.C., Abraham, J.A., Machwe, A.: User-centric evolutionary computing: melding human and machine capability to satisfy multiple criteria. In: Knowles, J., Corne, D., Deb, K., Chair, D.R. (eds.) Multiobjective Problem Solving from Nature. Natural Computing Series, pp. 263–283. Springer, Heidelberg (2008). https://doi.org/10.1007/978-3-540-72964-8_13
5. Cunha, J.M., Martins, P., Machado, P.: How shell and horn make a unicorn: experimenting with visual blending in emoji. In: Proceedings of the Ninth International Conference on Computational Creativity (2018)
6. Cunha, J.M., Martins, P., Machado, P.: Emojinating: representing concepts using emoji. In: Workshop Proceedings from ICCBR 2018 (2018)
7. Wicke, P.: Ideograms as semantic primes: emoji in computational linguistic creativity (2017)
8. Ha, D., Eck, D.: A neural representation of sketch drawings. arXiv preprint arXiv:1704.03477 (2017)

9. Karimi, P., Maher, M.L., Grace, K., Davis, N.: A computational model for visual conceptual blends. IBM J. Res. Dev. (2018)

10. Xiao, P., Linkola, S.: Vismantic: meaning-making with images. In: Proceedings of the Sixth International Conference on Computational Creativity (2015)

11. Correia, J., Martins, T., Martins, P., Machado, P.: X-faces: the exploit is out there. In: Proceedings of the Seventh International Conference on Computational Creativity (2016)

12. Pereira, F.C., Cardoso, A.: The boat-house visual blending experience. In: Proceedings of the Symposium for Creativity in Arts and Science of AISB 2002 (2002)

13. Cunha, J.M., Gonçalves, J., Martins, P., Machado, P., Cardoso, A.: A pig, an angel and a cactus walk into a blender: a descriptive approach to visual blending. In: Proceedings of the Eighth International Conference on Computational Creativity (2017)

14. Lourenço, N., Assunção, F., Maçãs, C., Machado, P.: EvoFashion: customising fashion through evolution. In: Correia, J., Ciesielski, V., Liapis, A. (eds.) EvoMUSART 2017. LNCS, vol. 10198, pp. 176–189. Springer, Cham (2017). https://doi.org/10.1007/978-3-319-55750-2_12

15. Rebelo, S., Fonseca, C.M.: Experiments in the development of typographical posters. In: 6th Conference on Computation, Communication, Aesthetics and X (2018)

16. Maçãs, C., Lourenço, N., Machado, P.: Interactive evolution of swarms for the visualisation of consumptions. In: ArtsIT 2018 (2018)

17. Dorris, N., Carnahan, B., Orsini, L., Kuntz, L.A.: Interactive evolutionary design of anthropomorphic symbols. In: Congress on Evolutionary Computation, CEC 2004, vol. 1, pp. 433–440. IEEE (2004)

18. Dozier, G., Carnahan, B., Seals, C., Kuntz, L.A., Fu, S.G.: An interactive distributed evolutionary algorithm (idea) for design. In: 2005 IEEE International Conference on Systems, Man and Cybernetics, vol. 1, pp. 418–422. IEEE (2005)

19. Piper, A.K.: Participatory design of warning symbols using distributed interactive evolutionary computation. Ph.D. thesis, Auburn University (2010)

20. Hiroyasu, T., Tanaka, M., Ito, F., Miki, M.: Discussion of a crossover method using a probabilistic model for interactive genetic algorithm. In: SCIS & ISIS SCIS & ISIS 2008. Japan Society for Fuzzy Theory and Intelligent Informatics (2008)

21. Gong, D., Yan, J., Zuo, G.: A review of gait optimization based on evolutionary computation. Appl. Comput. Intell. Soft Comput. **2010**, 12 (2010)

22. Pelikan, M., Goldberg, D.E., Lobo, F.G.: A survey of optimization by building and using probabilistic models. Comput. Optim. Appl. **21**, 5–20 (2002)

23. Wijeratne, S., Balasuriya, L., Sheth, A., Doran, D.: Emojinet: an open service and API for emoji sense discovery. In: Proceedings of ICWSM-17 (2017)

24. Speer, R., Havasi, C.: Representing general relational knowledge in conceptnet 5. In: LREC, pp. 3679–3686 (2012)

25. Browne, C.: A new general service list: the better mousetrap we've been looking for. Vocabulary Learn. Instr. **3**(1), 1–10 (2014)

26. Liapis, A., Yannakakis, G.N., Togelius, J.: Sentient sketchbook: computer-aided game level authoring. In: FDG, pp. 213–220 (2013)

27. Vinhas, A., Assunção, F., Correia, J., Ekárt, A., Machado, P.: Fitness and novelty in evolutionary art. In: Johnson, C., Ciesielski, V., Correia, J., Machado, P. (eds.) EvoMUSART 2016. LNCS, vol. 9596, pp. 225–240. Springer, Cham (2016). https://doi.org/10.1007/978-3-319-31008-4_16

28. Brysbaert, M., Warriner, A.B., Kuperman, V.: Concreteness ratings for 40 thousand generally known english word lemmas. Behav. Res. Methods **46**(3), 904–911 (2014)
29. Cunha, J.M., Martins, P., Machado, P.: Using image schemas in the visual representation of concepts. In: Proceedings of TriCoLore 2018. CEUR (2018)

Automatically Generating Engaging Presentation Slide Decks

Thomas Winters[1]([✉]) [iD] and Kory W. Mathewson[2]([✉]) [iD]

[1] KU Leuven, Leuven, Belgium
`thomas.winters@cs.kuleuven.be`
[2] University of Alberta, Edmonton, Alberta, Canada
`korymath@gmail.com`

Abstract. Talented public speakers have thousands of hours of practice. One means of improving public speaking skills is practice through improvisation, e.g. presenting an improvised presentation using an unseen slide deck. We present TEDRIC, a novel system capable of generating coherent slide decks based on a single topic suggestion. It combines semantic word webs with text and image data sources to create an engaging slide deck with an overarching theme. We found that audience members perceived the quality of improvised presentations using these generated slide decks to be on par with presentations using human created slide decks for the *Improvised TED Talk* performance format. TEDRIC is thus a valuable new creative tool for improvisers to perform with, and for anyone looking to improve their presentation skills.

Keywords: Computer-aided and computational creativity ·
Generation · Computational intelligence for human creativity

1 Introduction

Public speaking is difficult due to many psychosocial factors: 40% of surveyed adults feel anxious about speaking in front of an audience [1,2]. Practice and experience can make significant improvements to speech anxiety levels [3]. However, the overpracticing of a talk can make the delivery stiff and rehearsed. Improvisational theater has been shown to be effective in developing skills for complex social interactions [4,5]. Giving improvised presentations is thus a useful exercise for enhancing presentation skills and overcoming public speaking anxiety. The exercise itself is already performed by improvisational theater and comedy groups all around the world in several variants, such as in the form of an *"Improvised TED Talk"* [6]. This work presents the novel idea of automatically generating presentations suitable for such an improvisational speaking practice exercise, as well as for improvisational comedy performance.

One of the difficulties in the improvised presentation format is that the slide deck is created prior to performance, while in improvisational comedy, the audience provides suggestions to shape the show. The premade slide deck will thus

© Springer Nature Switzerland AG 2019
A. Ekárt et al. (Eds.): EvoMUSART 2019, LNCS 11453, pp. 127–141, 2019.
https://doi.org/10.1007/978-3-030-16667-0_9

not align with the suggestion, but instead just has *"curated random"* slides [6]. Another difficulty is learning how to build constructive slide decks for another performer to present. Creating such a slide deck requires time and attention, which can be costly and does not scale well when doing a large quantity of performances. To solve these problems, we present TEDRIC (*Talk Exercise Designer using Random Inspirational Combinations*), a co-creative generative system that is capable of creating engaging slide decks for improvised presentations. It is designed to fulfill a specific task of an improviser for a particular format, being the design of an improvised TED-talk slide deck [6]. In the improvised presentation format, one performer usually designs a slide deck for the other performer beforehand. Previous systems in this domain have experimented with fulfilling the presenter role, e.g. by writing text scripts, creating comedic performers, and all-round improvisational comedy agents [7–9]. TEDRIC on the other hand focuses on the task of the other performer, who creates the slide deck for the speaker. The system also differs from previous work in slide deck generation [10–12] in that rather than summarizing text into slides, or converting outlines to well-designed slides, it actively creates its own presentation story and chooses the content itself. This system is thus a good example of co-creation between humans and machines, as both have responsibility in the creative process of performing an improvised presentation. Thanks to the speed advantage TEDRIC has over its human counterpart, it can generate a slide deck based on a suggestion from a live audience on the spot. Another advantage TEDRIC has, is that it follows certain good design guidelines, something which individuals new to this format have struggled to learn quickly [6].

2   Slide Deck Generator Model

The TEDRIC system is composed of several linked generators, specified in a presentation schema. In this paper, we focus on the *Improvised TED talk* presentation schema, which is meant to give comedic suggestions to spark the presenter's creativity during the presentation. However, it is easy to create similar schemas for more serious presentations (e.g. an improvised *Pecha Kucha* [13]) by taking the subset of slide generators that do not use comedic content sources.

2.1   Schemas

The templates and schemas technique is a commonly used approach for generating content using certain constraints [14–17]. In this project, the presentation schema specifies a slide seed generator, a set of slide generators, slide generator weight functions and frequency limitations on slide generator types. The slide generators themselves also use a different kind of schema to generate their content. Such a slide generator schema specifies a set of content sources to fill placeholders in a specific PowerPoint slide template, resulting in a coherent slide.

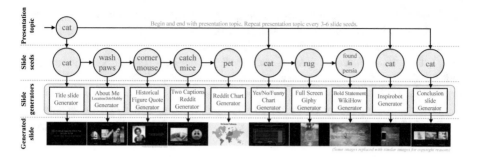

Fig. 1. TEDRIC system pipeline

2.2   Slide Seed Generator

A seed generator produces a list of slide seeds based on the overarching main presentation topic, provided by the audience. The default seed generator achieves this by randomly walking over links in ConceptNet, a semantic network of related words [18]. For example, it could find that *cat* is related to *rug* because cats can be found laying on rugs (Fig. 1). If no useful link is found for a certain slide, it backtracks and looks at the slide seed before the previous slide seed. The default seed generator constrains the first and last slide to be about the given presentation topic, as well as making this main topic return every three to six slides. It thus ensures that the given main topic is clearly present in the presentation, while still allowing some deviations to related topics.

2.3   Slide Generator

Slide generators use a given seed to generate a logical, themed slide for the presentation slide deck. During slide deck generation, the presentation schema iteratively picks one of its slide generators using their weight functions. These weight functions calculate the weight of a slide generator based on the current position of the slide to be generated and on how often similar slide generators can occur in one presentation. For example, the weight of the *"Title Slide Generator"* peaks for the first slide, and has a weight of zero everywhere else. Similarly, the several *"About Me"* and *"History"* slide generators prefer being in one of the first slides after the title slide, and thus have higher weights when choosing a slide generator for these positions. Each slide generator also has several associated tags. These tags are used to limit the number of certain slide types occurring in the same slide deck. For example, at most one anecdote slide is allowed, and no more than 20% of the total number of slides can be about a quote. For every slide, the presentation schema thus first creates a list of slide generators that can still be used given the slide generators used for other slides. It then calculates the weight for all these slide generators by giving the slide number and the total number of slides as arguments to their weight functions. It then picks the slide generator using roulette wheel selection on the weights of these slide generators.

After picking a slide generator, this generator then creates a slide based on a given seed, using its slide generation schema. To achieve this, it passes the seed to all content sources specified in this template. For example, using the seed *"rug"*, the *"Full Screen Giphy Generator"* passes this seed to Giphy to find a related gif, and to a title content source to generate a slide title using this seed word (Fig. 1).

The system currently contains 26 slide generators. They can be grouped in several different main categories based on the type of slide they produce, which are listed below.

Title Slide. The title slide generator creates a slide with a generated title about the main topic, based on a large set of templates. The seed is often transformed to a related action by using WikiHow (see Sect. 2.4). The slide also contains a subtitle with a generated presenter name (if none is provided) and a generated scientific sounding field (e.g. *Applied Cat Physics*) (Fig. 2a).

About Me Slide. There are several different slide generators introducing the speaker. They all generate texts and images visualizing the accompanied text, and can be about a location, a country, a job, a book and/or a hobby related to the presenter (Fig. 2b).

History Slide. The system contains several generators that make up history related to the given seed. The *"Historical Figure Quote"* picks an old picture (from Reddit), generates a name (using a context-free grammar), invents an achievement (using a WikiHow action and templates) and finds a related quote (using GoodReads) to create a fake historical figure related to the seed (Fig. 2c). There are also other slide generators using vintage pictures with captions relating to time. These history slides are used for adding depth and narrative to the generated presentation topic.

Full Screen Image Slide. Breaking up the slideshow with a full screen image is a great way for providing freedom as well as for grabbing attention. There are several variants of large image slide generators employed in TEDRIC, such as using Google Images to create a descriptive slide, where the speaker has a lot of freedom and breathing space to talk. Another variant uses a large animated gif and a title pointing the attention to the gif.

Statement Slide. Since the slide decks are meant for *Improvised TED talks*, we want them to feel as if they provide the audience with a take away message. To achieve this, we created several templates for bold statements, call to actions and anecdote prompts (Fig. 2e). The system also uses several online sources (such as InspiroBot and GoodReads) to provide inspirational quotes.

Multiple Captioned Images Slide. Some slide generators use multiple images on one slide, with randomly picked captions that belong together underneath (e.g. *"Expectation"* & *"Reality"*). We found that these slides work best if the last captioned image uses content source providing odd images, as they help the presenter with creating a strong punchline if done correctly.

Chart Slide. We implemented several ways of creating charts using generated data, as these usually give the presentation the perception of having more scientific foundation. One chart generator uses histograms and pie charts about yes/no questions. It generates a question about the slide seed using templates, and a funny third answer based on a large collection of textual templates (Fig. 2f). The data generator prefers generating high values for the funny answer. Similarly, ConceptNet provides a good source of related locations that can be used in a pie chart or histogram (Fig. 2d). Another type creates scatter plots by adding noise to basic mathematical functions (e.g. $y = x^2$ and $y = log(x)$). The labels of the axis are then found by using links in ConceptNet. A third type uses subreddits about data visualization for finding interesting related charts.

Conclusion Slide. Conclusion slides use a similar structure to the multiple captioned image slides. These slides usually start with a descriptive image about the main topic and have an odd image at the end, in order to finish the show with a punch.

2.4   Content Sources

Content sources are used in slide generators to provide them with either texts or images. The content sources produce this content based either on a suggestion or at random. Slide generators pass the seeds they receive from the presentation schema down to all of their content sources. Each content source then decides whether or not to use this seed. For example, if the content source uses a search engine, it might fall back to picking content at random in case the search engine does not return any results. Some content sources (e.g. a context-free grammar for generating full names) might not support using the seed at all, and thus always return completely random content. Ignoring the seed and picking content at random every once in a while can have a positive effect as it might provide more variety in a presentation. This is especially true when the slide seed generator is not able to find a lot of meaningful related words, e.g. for an uncommon, niche word.

The system uses two types of content sources: image sources and text sources. These use search systems, context-free grammars, templates and combinations of the three. One advantage search systems have over pure context-free grammars and templates is that they are usually larger, contain more diverse content and are frequently updated. This gives rise to more surprises to both the creators as well as the users of the system, as new entries are usually constantly being added to these sources, generating different content every time it is used. Since such

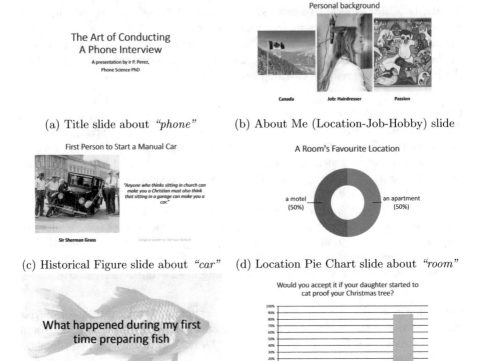

Fig. 2. Possible outcome examples of slide generators

sources are harder to predict, it is important to identify the style and flavour of the elements in the search system or corpus in order to successfully use it in a slide generator schema. For example, stock photo websites usually give more neutral images, whereas certain subreddits might give funny or cute images. Sources of similar flavour can then be combined into composite content sources, to increase the variety a slide generator can produce. For example, the Google Images data source can be combined with the stock photo data source, as both provide neutral images. The slide generators then use these content sources in places where they need data in such a specific style and flavour.

Image Sources. The system uses a large variation of image sources. In order to find descriptive pictures, it uses Google Images and stock photo websites. The system also uses images from Inspirobot[1], a bot that generates nonsensical inspirational quotes, and frames them using a thought-provoking background.

---

[1] http://inspirobot.me/.

We also built our own data generation system for the analytical slide generators. This data source is used to insert random data into charts using labels created by text sources. Reddit[2] is also a great source of images, as a lot of its subreddits usually contain images in a particular style or flavour. Images from Reddit are usually intrinsically interesting, as users tend to have good reasons for thinking the picture is worth posting on Reddit. The number of upvotes for a Reddit post generally gives a good indication for the quality of the image as well. We specifically use subreddits about odd, punchline images (e.g. *r/hmmm*, *r/wtfstockphotos*, *r/eyebleach*), gifs (e.g. *r/gifs*, *r/nonononoYES*), historic and vintage pictures (e.g. *r/OldSchoolCool*, *r/TheWayWeWere*, *r/ColorizedHistory*), charts and other data visualisations (e.g. *r/dataisbeautiful*, *r/funnycharts*), book covers and outside pictures. The Reddit gif source is combined with the Giphy[3] source to create a composite gif data source.

Text Sources. We employed several types of text sources in this project. The first type utilizes search engines to find texts related to a slide seed. As such, the system uses GoodReads[4] as a source for related quotes, to be used on slides about historic people or for large statements. Another such search engine is WikiHow[5], which we use to find human actions related to a certain seed, or related to other actions. This is achieved by searching on WikiHow for the given input and looking at the titles of the found articles. Considering WikiHow article titles follow structures such as *"How to ..."* or *"5 ways to ..."*, it is trivial to extract actions related to the input from these. For example, searching for *"cat"* on WikiHow finds articles such as *"How to Pet a Cat"*, which will lead to the action *"to pet a cat"*, which can then for example be inserted into a template (e.g. "Why We All Need to {seed.wikihow_action.title}") to form the talk title *"Why We All Need to Pet a Cat"*.

For creating controllable text content sources with high variation, we built our own text generation language on top of Tracery, a text generation language using context-free grammars [19]. The extensions allows us to specify external variables with custom functions in the template texts. These custom functions can be used on the externally specified variables, and thus transform them into right forms, or chain other functions. They can also be used as a predicate (e.g. checking if the variable is a noun). The transformation functions are used to change the form of the variable (e.g. verb conjugation, pronouns or string casing) or finding related texts (e.g. related actions using WikiHow or related concepts using ConceptNet [18]). This extension is implemented by first making Tracery generate text which might have external variables and functions specified between curly brackets. After this generation, the system checks if there are missing external variables, failed predicate checks on these variables and if they are transformable using the specified functions. If all conditions are satisfied, the

---

[2] https://reddit.com.
[3] https://giphy.com/.
[4] https://www.goodreads.com/.
[5] https://wikihow.com.

variable is filled in, transformed and inserted in the textual template. Otherwise, the generator keeps trying until all conditions are met. The extended generative grammar language allows us to specify a large number of generators, such as generators for historic people names, for bold and inspirational statements, talk (sub)titles, related sciences, slide title variations and punchline chart question answers.

Tupled Sources. For specifying certain slide generators, we required tupled data sources, i.e. data sources that inspire each other, or that are generated at the same time. For example, when generating a job, we want to generate an image based on the generated job function. When generating slides with two captions, these captions should sometimes also relate to each other (e.g. *"good"* & *"bad"*, *"what I initially did"* & *"actual solution"*). For this, the system thus also allows the generation of content using tupled data source.

2.5   Parallelisation

Since the slide decks are meant to be generated in real-time in front of an audience, we need the generator to finish within a reasonable time frame. To achieve this, the generator has the option of generating all slides at once in parallel. This decreases the influence of slow (e.g. online, rate-limited) data sources. However, naively generating all slides in parallel causes issues with the validity of the slide deck, as the generator has several constraints that must hold between generated slides, such as limited duplicate images and restrictions on maximum tag occurrences of the used slide generators. To solve this, TEDRIC generates the slide seeds and selects some slide generators. It then fires all the slide generators at once without giving them any knowledge about other slides, since there are no other slides yet. After all initial slides are generated, TEDRIC sequentially checks every slide, increasing its saved knowledge about the slides after viewing every next slide. If a slide is breaking any of the constraints, its slide number is stored for the next generation round. After viewing all slides, a new slide generation round is fired, but this time with knowledge from all valid, generated slides. This repeats itself until no new slide breaks any constraints. At the end of this process, the generated slides respect the constraints specified by the presentation schema similar to slides generated in a serial way.

2.6   Code

The code, documentation and link to an online demo of TEDRIC is available on https://korymath.github.io/talk-generator/.

3   Evaluation

We evaluated the system in several different ways. First, we made volunteers present improvised presentations and formally give feedback to all presentations.

This feedback is then analysed to compare the quality of human created slide decks versus generated slide decks. Second, we evaluated the time performance of the full slide generation process in a context similar to the expected use case context, to evaluate if it completes within a reasonable time frame. Third, we informally evaluated the system by training some performers using generated slide decks, and making them perform the format on stage using slide decks generated during the performance.

3.1 Quantitative Results

To evaluate the quality of our system, we compared the perceived quality of generated slide decks with slide decks created by humans by performing them in the improvised presentation format with eight volunteers. In total, six talks were given. We generated three slide decks on the spot based on the audience suggestion for the presentation topic, and used three slide decks created by three different people that were used before to perform this format on stage [6]. These six presentations were presented in mixed order, so that the participants did not realize that there were two different sets of presentation sources. Participants had to indicate whether they were presenting or listening to the presentation, and had to give two ratings, one for the given presentation as a whole, and one for the slide deck quality. We used a five-level scale, with five being the highest score (*Awful, Bad, Average, Good, Amazing*). A disadvantage of this evaluation is that the performative nature might be a confounding variable. However, this might also help the participants assess the quality of the slide deck better than rating the slide deck in separation.

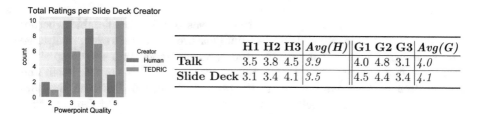

	H1	H2	H3	Avg(H)	G1	G2	G3	Avg(G)
Talk	3.5	3.8	4.5	*3.9*	4.0	4.8	3.1	*4.0*
Slide Deck	3.1	3.4	4.1	*3.5*	4.5	4.4	3.4	*4.1*

Fig. 3. Comparing the (H)uman created and (G)enerated slide decks, for both the quality of the talks themselves and the slide deck quality (N = 8, for all six presentations)

With the gathered data[6] on Fig. 3, we can check if the source of the slide deck is correlated with the quality of the slide deck. When performing a t-test on the dataset to see if the distributions are different, we find $p = 0.039$. This value allows us to reject, within the 5% confidence interval, the hypothesis that they

---

[6] The evaluation data is made available on https://github.com/korymath/tedric-analysis.

have the same distribution. However, when performing a χ^2 test on contingency table without scores of 1 (as none were present), then $p = 0.148$, which is not significant enough to reject that they have a similar distribution. This is likely due to the fact that there were only 48 observations. Either way, the generated slides generally do not seem score lower in our sample than the human created slide decks.

3.2   Qualitative Results

The feedback forms also had a section where the participants could fill in short, textual comments about their given slide decks, of which they did not know the source. There were 30 comments in total, which we described in a thematic analysis using codes that describe whether they are positive (POS) or negative (NEG) about the human created slide deck (H), generated slide deck (G), theme and presenter as well as common reasons for these sentiments (Fig. 4).

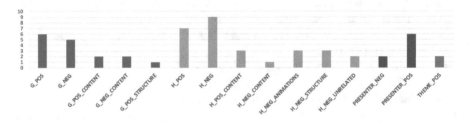

Fig. 4. Codes and their frequency f (if $f > 0$) in the thematic analysis

Some presenters had complained that it was quite unpredictable whether their given human-created slide decks would feature a new slide or an additional animation next, making it harder to present. Some participants also complained that these human-created slide deck had nothing to do with the talk itself, and that it often put too many words in the presenter's mouth. However, they did enjoy that slide decks had clever jokes, nice variation and backreferences with returning images.

Some participants complimented the generated slides on their nice structure and on the ease to link them to the presentation topic. The participants wanted to try a variation after seeing five talks, and proposed to test out the format in a panel show setting, with one moderator and two guests. The generated slide deck (*G3* in Fig. 3) did not support this type of talk well, as it was too focused on one person (e.g. in the *"About Me"* slide), giving rise to some negative comments about the content. The amount of text was too high, as there were three improvisers on stage, who all wanted to add new information as well. In order to support this type of presentation exercise well in the future, we will thus need an additional, revised version of the presentation schema.

3.3   Time Performance

Since our generator is used in front of a live audience, it should be able generate within a reasonable time frame. To optimally estimate the timings of the typical use case of this generator, we mimicked such a situation as well as possible. Firstly, we used 500 of the most common English nouns[7] as presentation topics for the generated slide decks, as these fairly accurately represent audience suggestions. Choosing more common nouns as topics are also a worst-case estimate for time performance. This is due to less common nouns having less relations, which increase the odds of slide topic generator cache hits, which also leads to more similar slide topics, increasing the probabilities of hitting content generator caches. Secondly, we ran the timing experiment on a normal laptop with download speeds fluctuating between 80–150 megabits per second, as this is similar to the set-up a goal user might have. Thirdly, the slide decks were not only represented in-memory, but also saved to a PowerPoint file. Fourthly, we chose to generate slide decks of seven slides, as this is a nice length used for practising the format, as well as for presenting snappy, humorous presentations. Increasing this length should not increase the generation time with a similar factor due to the fact that the slides are generated in parallel rounds.

We can see on Fig. 5 that 94.4% of slide decks are generated within a 20-second time window, with a median of 12.29 s, and a maximum of 38 s (and one outlier of 58 s). The time spent on actual computations in the program has a median of 2.49 s, meaning that about 80% of the total generation time is spent waiting on the response of remote content sources. Better internet connections can thus decrease the generation time even further. We can conclude from this experiment that the generation time of TEDRIC is low enough to be used in presentation workshops and on stage.

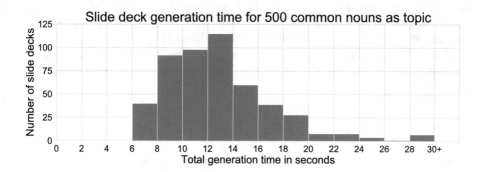

Fig. 5. Generation time of slide decks with seven slides

---

[7] https://www.wordexample.com/list/most-common-nouns-english.

3.4   Experiences on Exercise and on Stage Performance

We organized a small workshop and coached a group of four improvisational theater actors using this generator. They quickly improved their presenting methods. The narrative of their stories initially lacked logic and did not link back to
the generated title of the talk, or convinced the audience about the importance
of the topic. They also rarely dared using a bold statement right before going
to the next slide, a technique for making the audience laugh (e.g. *"The next
slide will provide the solution to this problem."*). These skills were easily trained
using the generator, and at the end of a three hour workshop, all participants
gave talks that avoided these mistakes. Later that week, they were put on stage
during a comedy festival, where people could leave at any time to see other
comedy acts. They managed to keep all thirty to forty people interested enough
to stay seated during all four talks. Furthermore, they were able to make the
audience laugh during almost every slide, thanks to the inspiring slide decks. We
see this as a type of collaborative creativity between humans and computers.
However, one thing we noticed was that the energy of the audience dropped
when the fourth talk began. A likely reason for this has been described in the
past as an *"ice cream after an ice cream situation"* [6]. One way of overcoming
this problem in the future would be to implement more types of presentation
schemas, which could be achieved by slightly varying the current presentation
schema (e.g. presentation schemas preferring more art, science or nature slides).

3.5   Interpretation

In the evaluation sample, the generated slide decks score higher on average than
the human-created ones, and make the presenters give better presentations overall (Fig. 3). One explanation for this is that the generated presentations are more
related to the given audience suggestions, whereas the premade slide decks are
not. This results in a more coherent presentation. Even though this is a small
evaluation sample, it is a good indication that the system is performing well and
is usable for enhancing the performance of presenters. We also showed that the
generator is valuable and usable for coaching improvisers and even works well
on stage.

4   Future Work

There are several extensions we propose for future work. Obvious extensions
are adding more variation to TEDRIC by adding more types of slide generators,
different flavoured presentation schemas and slide deck themes. We hope that
making this tool available for other performers might also incentivize more tech-
savvy users to add their new slide generators to the code repository, giving rise
to such additional variations.

This research also opens the path to using genetic algorithms for more
human-curated co-creation of slides. This could be achieved by iteratively

proposing several slides to the user, and giving the selected slide(s) higher fitness values. A slide cross-over operator could use the slide generator to swap the content filled in certain placeholders with content from another slide. Similarly, a mutation operator could just regenerate content in a certain position by calling this part of the generator again. A user would then be able to make slides evolve using human curation, while the system keeps its important role in the creative aspects. In this paper, we mainly put our focus on autonomous seeded slide generation, as the goal was to achieve real-time generation during performances.

One more ambitious goal of this research is to build an end-to-end system which can, in real-time, iteratively generate a slide deck using interaction-based machine learning models [8,17,20]. This means that every next slide could be adapted to the response of the audience to previous slides, in order to further optimize engagement. For these types of sequence generative models, the training data set must contain many examples of rich, contextual, structured pairs. In order to generate such interactive presentations with e.g. neural models, we must develop an understanding of how to automatically generate a presentation comparable to a human using rules and heuristics, as we started in this paper. We hope that our tool is used by performers who see the advantages of our system, as such a community could help us build a data set to open the path to a more interactive slide deck generation system.

5   Conclusion

In this paper, we presented a system for generating engaging slide decks called TEDRIC. This system can be used for both training presentation skills as well as for performing improvisational comedy in the *Improvised TED Talk* format.

There are several advantages our system has over humans having to create slide decks. Firstly, it has the advantage of being able to create slide decks about an overarching theme decided by an audience suggestion. We showed in our evaluation that this might increase the perceived quality of the given presentation. Secondly, it performs the task of creating slide decks for this format several orders of magnitude faster than humans can. Thirdly, it significantly lowers the barrier to perform improvised presentations. Not only is it able to follow good guidelines, something that performers often have struggled to learn quickly before, it also relieves the performers of the time consuming task to create slide decks for each other to practise.

The behaviour of the presented system is a type of co-creation between computers and humans, as it provides surprising slides to spark the creativity of the presenter, similar to how humans creating a slide deck would. We thus hope that this system will not only improve the presentation skills of humans, but also make them more acceptant towards computers aiding their creativity.

Acknowledgments. Thank you to the reviewers for their time and attention reviewing this work, as well as for the insightful comments. Thanks to Julian Faid and

Dr. Piotr Mirowski for advice and support in the creation of the software. Thank you to Lana Cuthbertson, the producer of TEDxRFT, and to the talented individuals at Rapid Fire Theatre for supporting innovative art. Thank you to all the performers who have done improvised TED talks and shared their views and opinions on how to structure and frame, your help in building the design guide was critical. Thanks to Shaun Farrugia for creating and hosting an online demo of TEDRIC, making it more easily available for performers of any background to use the system. Thank you to volunteers from the Belgian improvisational comedy group Preparee for volunteering for the evaluation.

References

1. Wilbur, P.K.: Stand up, Speak up, or Shut up. Dembner Books, New York (1981)
2. Dwyer, K.K., Davidson, M.M.: Is public speaking really more feared than death? Commun. Res. Rep. **29**(2), 99–107 (2012)
3. Lee, S.: Study on the classification of speech anxiety using q-methodology analysis. Adv. Journalism Commun. **2**(03), 69 (2014)
4. Watson, K.: Perspective: Serious play: teaching medical skills with improvisational theater techniques. Acad. Med. **86**(10), 1260–1265 (2011)
5. King, B.T., Janis, I.L.: Comparison of the effectiveness of improvised versus non-improvised role-playing in producing opinion changes. Hum. Relat. **9**(2), 177–186 (1956)
6. Mathewson, K.W.: Improvised ted talks (2018). https://korymathewson.com/improvised-ted-talks/
7. Winiger, S.: Ted-RNN – machine generated ted-talks (2015). https://medium.com/@samim/ted-rnn-machine-generated-ted-talks-3dd682b894c0
8. Knight, H.: Silicon-based comedy (2010). https://www.ted.com/talks/heather_knight_silicon_based_comedy
9. Mathewson, K.W., Mirowski, P.: Improvised theatre alongside artificial intelligences. In: AAAI Conference on Artificial Intelligence and Interactive Digital Entertainment (2017)
10. Hu, Y., Wan, X.: PPSGen: learning-based presentation slides generation for academic papers. IEEE Trans. Knowl. Data Eng. **27**(4), 1085–1097 (2015)
11. Ranjan, A., Gangadhare, A., Shinde, S.V.: A review on learning based automatic PPT generation using machine learning. Int. J. Sci. Res. Comput. Sci. Eng. Inf. Technol. (IJSRCSEIT) **3**, 547–552 (2018)
12. Sravanthi, M., Chowdary, C.R., Kumar, P.S.: SlidesGen: automatic generation of presentation slides for a technical paper using summarization. In: FLAIRS Conference (2009)
13. Beyer, A.M.: Improving student presentations: Pecha Kucha and Just Plain PowerPoint. Teach. Psychol. **38**(2), 122–126 (2011)
14. Binsted, K., Ritchie, G.: An implemented model of punning riddles. CoRR abs/cmp-lg/9406022 (1994)
15. Manurung, R., Ritchie, G., Pain, H., Waller, A., Mara, D., Black, R.: The construction of a pun generator for language skills development. Appl. Artif. Intell. **22**(9), 841–869 (2008)
16. Venour, C.: The computational generation of a class of puns. Master's thesis, Queen's University, Kingston, Ontario (1999)

17. Winters, T., Nys, V., De Schreye, D.: Automatic joke generation: learning humor from examples. In: Streitz, N., Konomi, S. (eds.) DAPI 2018. LNCS, vol. 10922, pp. 360–377. Springer, Cham (2018). https://doi.org/10.1007/978-3-319-91131-1_28

18. Liu, H., Singh, P.: ConceptNet: a practical commonsense reasoning tool-kit. BT Technol. J. **22**(4), 211–226 (2004)

19. Compton, K., Kybartas, B., Mateas, M.: Tracery: an author-focused generative text tool. In: Schoenau-Fog, H., Bruni, L.E., Louchart, S., Baceviciute, S. (eds.) ICIDS 2015. LNCS, vol. 9445, pp. 154–161. Springer, Cham (2015). https://doi.org/10.1007/978-3-319-27036-4_14

20. Jaques, N., Engel, J., Ha, D., Bertsch, F., Picard, R., Eck, D.: Learning via social awareness: improving sketch representations with facial feedback. arXiv preprint arXiv:1802.04877 (2018)

Tired of Choosing? Just Add Structure and Virtual Reality

Edward Easton$^{(\boxtimes)}$, Ulysses Bernardet, and Aniko Ekart

Department of Computer Science, Aston University, Birmingham, UK
{eastonew,u.bernardet,a.ekart}@aston.ac.uk

Abstract. Interactive Evolutionary Computation (IEC) systems often suffer from users only performing a small number of generations, a phenomenon known as user fatigue. This is one of the main hindrances to these systems generating complex and aesthetically pleasing pieces of art. This paper presents two novel approaches to addressing the issue by improving user engagement, firstly through using Virtual Environments and secondly improving the predictability of the generated images using a well-defined structure and giving the user more control. To establish their efficacy, the concepts are applied to a series of prototype systems. Our results show that the approaches are effective to some degree. We propose alterations to further improve their implementation in future systems.

Keywords: User fatigue · Interactive Evolutionary Computation · Virtual Reality

1 Introduction

Creativity is a part of life that has existed for as long as humankind has existed, from cave paintings to modern-day graffiti artists, it runs deep throughout our very existence [1]. Integral to this is the ability of humans to determine the aesthetic quality of different works of art. Within Evolutionary Art one of the major difficulties is how to effectively formalise these ingrained notions of aesthetics into an automated fitness function. Many approaches have been used to address this problem such as semi-automated systems [2] or fully manual systems [3]. Both types of system try to overcome the problem by replacing, at least partially, the automated fitness function with a human user. These systems are termed Interactive Evolutionary Computation (IEC) systems [3] and this approach has proved effective in a range of different applications such as fashion [4]. Issues exist with both types of system: semi-automated systems still have problems with being able to define user preferences in terms of attributes of the image and fully manual systems are limited by the amount of input received by the user, a limit which is known as user fatigue [3]. The disparity between the number of generations which can be processed between the two types of system

© Springer Nature Switzerland AG 2019
A. Ekárt et al. (Eds.): EvoMUSART 2019, LNCS 11453, pp. 142–155, 2019.
https://doi.org/10.1007/978-3-030-16667-0_10

is quite high, semi-automated systems can easily complete 60 or more generations [5], whereas the user fatigue limit for fully manual IEC systems is known to be around 10–20 generations. Due to this, user fatigue is a major reason for IEC systems not being able to generate suitably pleasing images. Many novel methods have been introduced in recent years to overcome the limitations enforced by user fatigue such as branching [6], however these methods often enforce other requirements on the system, such as having access to multiple users.

User performance is not a new phenomenon within Computer Science and has often been studied in other application types, often being linked to the concept of user engagement [7]. Within IEC systems, there are two areas where the engagement can be improved.

One of the reasons cited in IEC systems for the presence of user fatigue, is the user is presented with options which do not appear to be related to their previous choices [8]. This can cause the user to stop being engaged with the system and ultimately stop completing generations.

Many IEC systems start with a simplistic, random image and expect the user to provide complexity and a direction to it. This may prevent users from starting to use the application or to quit within the first few generations as the images presented do not engage them sufficiently.

From research conducted in other fields, many different methods of improving user engagement have been suggested such as gamification [9]. One industry which continually encourages users to use their applications for long periods of time and perform a very high number of actions in a single session is the video games industry. This can be seen with users reportedly using the applications for up to 13 h a week on average [10]. By using techniques to improve the user engagement of the system, we can potentially increase the user fatigue limit. This has already been seen in existing contexts such as the Galactic Arms Race game [11] or Petalz [12] where evolutionary algorithms were applied to content generation which helped to engage users as much as possible.

2   Proposed Approaches

User engagement is addressed directly within this paper, which introduces two approaches designed to engage users by providing a more predictable experience and presenting artwork to the user in interesting ways. In order to provide a more predictable experience for the user a set structure is described, which the system can follow to present more appropriate images to the user and give them greater control over what is being changed. To give the user new and exciting ways to view the generated artwork, an immersive Virtual Reality environment is presented. These two approaches can be used in isolation and combination to encourage all users to use the system for longer and perform more generations. This section contains details about these approaches. Section 3 looks at how these approaches were realised into a practical application, Sect. 4 contains the experiment design and major results found from evaluating these strategies and finally Sect. 5 presents the conclusions and future work related to these concepts. Full details on the project can be found in [13].

2.1   Engagement Through Virtual Reality

A major reason to look towards the computer gaming industry when searching for methods to improve user engagement, is that IEC applications share a fair amount of commonality with gaming applications; they both offer a goal for a human user; they both rely on user input to achieve these goals; they both present an interface and control methods for the user to provide input to them (Fig. 1).

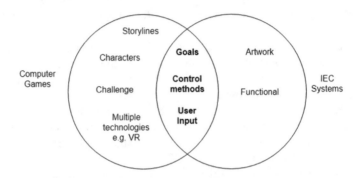

Fig. 1. Commonality between gaming and Interactive Evolutionary Computation (IEC) applications, including the goals, controls and user input aspects which feature in both system types

The gaming industry excels at adopting new technologies and applying them to existing applications. An example of this is the adoption of Virtual Reality (VR) through Head Mounted Displays (HMDs) such as the HTC Vive [14]. Between the market introduction of the first version of the headset in April 2016 and September 2018, over 2500 games have been released for VR on the Steam platform [15]. Though the use of the headset introduces some negative aspects such as its cumbersome nature and its weight, the game industry still offers engaging experiences for VR. This is achieved by utilising the headsets and the unique opportunity VR offers to provide the user with new and exciting ways to experience activities.

There are many aspects which contribute to the user engagement in computer games, such as the storyline and the characters. The application presented in this paper concentrates on a single aspect, the visuals, as a method of improving the user engagement with these systems. This is achieved through the use of VR and presenting the user with images in a 3D context which they can walk around and inspect from multiple different angles. With this added incentive over other evolutionary art applications, we can give the user a reason to want to start and to continue using the system.

2.2 Improving Predictability

The lack of predictability within IEC systems can be caused by two aspects, the first is the limited amount of information provided to the system by the user. In most IEC applications the user can only select a small number of preferences and from this information the system cannot determine which aspects of an image the user likes. In addition to this, common data structures such as expression trees have a very low locality [16], where small changes in the representation can lead to large changes in the displayed phenotype. This makes the process of relating elements of the data structure to elements of the produced image very difficult. There are multiple ways which this can be addressed, one method is to use high locality data structures such as the ones presented by Re [17]. However, in doing so the benefits and effects of using other data types are lost.

This paper introduces a set structure to an IEC system in order to improve on the issues mentioned above, whilst still allowing the application to use low locality data structures like expression trees. This gives the user greater control which ultimately leads to the images being amended more predictably for them. The systems are split into 3 different modes, these modes all perform different functions in the process of generating an image. Each one presents multiple images to the user but internally uses different evolutionary processes to change these images over every generation. These three stages are the rough tune mode, the fine tune mode and the lock mode, the relationships between the modes are shown in Fig. 2.

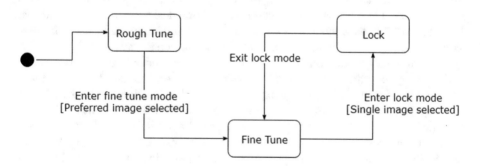

Fig. 2. State diagram of the transition between modes.

Rough Tune Mode. The rough tune mode is intended to allow the user to be able to generate a suitable base from which they would like to continue. At this stage the system makes coarse changes to the images without worrying about the outcome. This allows the user to quickly find a base image they like without having to believe it is the best possible image they can create through the system. The user can give a slight direction to the system by selecting their preferred option. Once the user has found an image they think has potential, then they can enter the fine tune mode.

Fine Tune Mode. The fine tune mode is designed to take the preferred image from the rough tune mode and then make minor adjustments to move it closer towards being the most aesthetically pleasing picture the user believes is possible. Depending on the implementation, this mode can introduce the use of data structures which have a high locality allowing a greater connection between the algorithms used and the user. The mode should be designed so that the user can see the relationship between subsequent sets of images.

Lock Mode. The final mode introduced is the lock mode, this mode gives the user a high level of control over what is being changed behind the scenes. It allows the user to select specific regions of a presented image and mark them as not needing any further changes i.e. locked. The system then takes this information and ensures that no further processing will take place on the specified region, until the user manually unlocks it. This mode helps to ensure the system continues to amend images in alignment with the user's intentions, without having to implement complex processes to automatically determine this information. In addition to this, the effective locking of regions within the image effectively reduces the search space the algorithm needs to work through to find the best image for the user.

3 Practical Realisation of the Proposed Approaches

To demonstrate the effect these suggestions can have, a series of three systems were created. These systems were comprised of one desktop application and two VR applications, each system being used to represent different levels of immersion. The desktop application represented a no immersion system, the first VR system offered a partially immersive experience being rendered at a low resolution and offering no auditory or haptic feedback to the user. The last VR system was designed to offer a fully immersive experience at high resolution and both auditory and haptic feedback provided. All systems implemented the different modes proposed in this paper, the effectiveness of these can be measured by the comparison of generations completed and the time spent whilst within each system, alongside user feedback. The images were generated using a particle image system similar to ones which has been used in previous applications [18, 19].

3.1 Environment Design

Desktop System. The desktop system was designed with simplicity in mind, where for each mode the user would see a single form displaying the current images. The system was interacted with using standard desktop controls of context menus and buttons, these allowed the user to select their preferred images, complete generations and lock particles. The layouts of the forms are shown in Fig. 3.

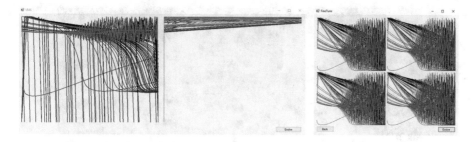

Fig. 3. Rough Tune and Fine Tune forms in desktop application

The locking of the particles within the desktop environment took the form of allowing the user to cycle through each of the individual particles which combined to construct the image and mark them as being locked or unlocked, as shown in Fig. 4.

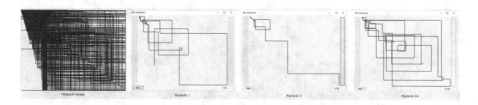

Fig. 4. Desktop locking forms

Virtual Reality Systems. To help the user feel a sense of familiarity with the Virtual Environment, it was based around the metaphor of a "smart art gallery". The user was placed in a large, open, minimalistic room and the images were presented within frames hanging on the walls. The system used animation to cover the loading times for completing generations, rendering the pictures and to indicate that the gallery itself was amending the images behind the scenes. To allow the system to render the images appropriately each set of items displayed to the user would move around the room in an anti-clockwise direction, where the next set of images could be rendered without interfering with existing ones. Other minor objects were introduced into the environment to help further invoke the sense the user was in an art gallery such as placing a bench in the middle of the room and having small plaques on the left side of the frames. An overview of the environment is shown in Fig. 5.

In order to present the artwork in new and exciting ways, the VR systems implemented three aspects: the first was allowing the user to move around the environment through teleportation, this let the user easily view the artwork from multiple different angles and from different distances; the system implemented a different version of the lock mode to the desktop environment, where upon

Fig. 5. Art gallery environment

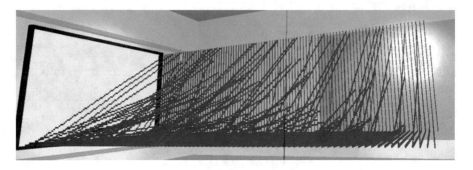

Fig. 6. Inspecting the individual particles in VR

entering the mode, the user was presented with a deconstructed view of the particle image, showing each individual particle, as shown in Fig. 6; finally the system gave the user the ability to take full control over the generated artwork by exporting them as a high quality model into Google's Tilt Brush software [20], an example of a particle image rendered in Tilt Brush is shown in Fig. 7. Further details about the virtual environments and the Tilt Brush integration can be found at [21].

3.2   Genetic Algorithm

The algorithms in use for the particle systems followed the standard genetic algorithm pattern, using a series of expression trees as the data representation of each image. This algorithm was used in all modes of the systems, however, to make coarser changes to the image within the rough tune mode, only mutation was used. After each generation was completed the entire population was replaced by the children created through the mutation and crossover operations.

Fig. 7. Tilt Brush rendered particle model

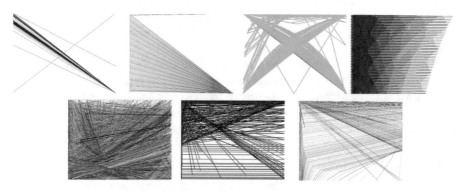

Fig. 8. Example generated images

Each image was defined as a series of nine expression trees, three were used to determine the initial X, Y, Z position of the particle; three were used to determine the initial RGB colour of the particle and the final three were used to calculate the X, Y, Z position of the 100 particles over a series of 100 time steps. Due to the integration with Tilt Brush only the X, Y and Z components could be updated and so the entirety of a particle was kept the same colour. Examples of items generated using the particle system are shown in Fig. 8.

One of three actions could be performed as part of the mutation on a tree, the first action is to entirely replace the node with a new expression but keeping the same child nodes, the second action is to remove a node and all its children entirely and finally just the children of a node could be replaced.

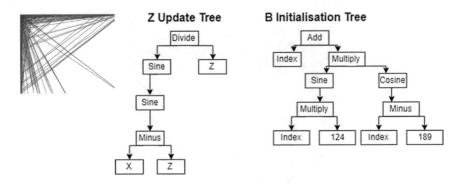

Fig. 9. Example particle image and expression trees

Within the fine tune mode, single point crossover was used by selecting a node at random from both selected parents, the first child was generated by using the tree from the first parent with the selected node replaced by the one from the second parent and the second child is generated by using the tree from the second parent with the selected node replaced by the one from the first parent. This crossover was applied to each of the 9 trees on each individual (Fig. 9).

The use of expression trees as the data structure did cause problems when implementing the locking due to the low locality of the representation. To compensate for this the system took a copy of the 9 expression trees used to generate the locked particle and proceeded to use these trees to plot the path of the particle in all future generations whilst the particle was marked as locked (Tables 1 and 2).

Table 1. Parameters used for the rough tune mode

Representation	9 expression trees
Population size	2
Mutation operator	0.2 (preferred item), 0.4 (non-preferred item)

Table 2. Parameters used for the fine tune mode

Representation	9 expression trees
Population size	4
Crossover operator	Single point crossover
Mutation operator	0.2

4   Experiment and Results

4.1   Experiment Design

To investigate the potential effectiveness of the introduced concepts and the created systems, we conducted a small pilot study. Two female and one male participant were given up to 15 min to use each of the three systems, being tasked to create the most aesthetically pleasing image they could and to only use the system for as long as they wished to. After the participant had finished using each system, they were asked to complete a short questionnaire, that measured aspects such as their feeling of presence within the system; how effective they believed the system to be and how much they enjoyed using the system. In addition to the questionnaire, the system also recorded user actions, including the mode changes made, the number of generations which were completed, the time spent in each system and how many particles were locked and unlocked.

Within the three systems created, it would be expected that the highest values for user satisfaction and number of generations completed are found on the full immersion VR system and the lowest on the desktop system. This will be a clear indication that all the proposed methods in this report contribute effectively to reducing user fatigue.

4.2   Results

The analysis of the usage duration shows that the full fidelity system was being used for almost twice as long as the desktop system (Table 3).

The questionnaire data showed that the users rated the full immersion VR system higher, both in terms of how much they enjoyed the interaction, and as to whether the participant would choose to use the system more in future (Fig. 10).

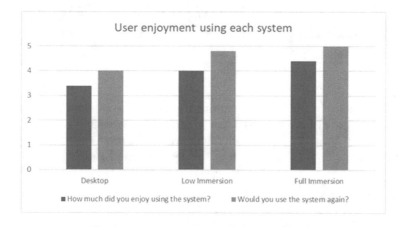

Fig. 10. User ratings for enjoyment out of 5

The ratings showed that all of the systems consistently allowed the participant to feel that they were able to make noticeable progress towards generating an aesthetically pleasing image (Fig. 11). When asked how excited they were about seeing the next set of images, participants rated the VR higher than the other systems. This indicates that the VR systems did add value to the user and improve engagement.

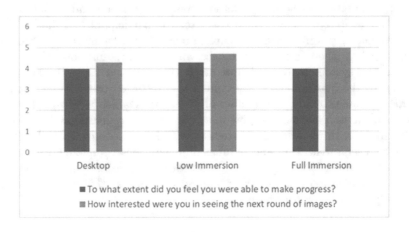

Fig. 11. User ratings for ability to make progress, marked out of 5

However, in contrast to these findings, the actual number of generations being completed using the VR system were relatively low, compared to a much higher number completed within the desktop system (Table 3).

Table 3. Basic usage statistics of the three systems

	Desktop	Low immersion	Full immersion
Avg. time spent [m:s]	6:03	8:29	10:03
Avg. generations completed [#]	35	9.3	8.3

User Feedback. In addition to quantitative results, we received a high amount of positive informal feedback from the participants, especially surrounding the Virtual Reality systems and the lock mode, where the users enjoyed being in the lock mode and viewing the artwork in the deconstructed view.

5   Conclusion

The results obtained provide a mixed indication about the systems being tested. Overall each system seemed to work as intended and generally the use of VR and multiple stages within the evolutionary process both improved on the user engagement and predictability of the presented images.

The use of VR seemed to provide three benefits; the user spent longer within the systems, enjoyed using them more and were more interested in seeing the next round of images, all three of these benefits indicate the users were more engaged with the system. However, this did not lead to an increase in the number of generations completed as shown in Table 3. There could be many different reasons for this such as the small size of the pilot study, however in the authors' opinion the implementation of the environments could have been the main cause of this. The virtual environments gave the user a greater set of available activities such as inspecting the images from multiple angles or spending an increased amount of time in the lock mode in order to view the deconstructed image, other aspects which may have reduced the number of generations completed could include the time it took for the animations to show and hide the frames. These aspects of the implementation would all have contributed to severely reducing the number of generations the user could perform in the VR environments and would explain the results we obtained.

The consistency of the rating the users gave all systems for ability to make progress suggests that the inclusion of the three modes in the system positively impacted its ability to provide predictable images to the user. This did not translate into an increase in the number of generations performed, however given the limitations mentioned above, this is perhaps not surprising.

A lot of positive points can be taken from the work presented in this paper, both the use of VR and modes seemed to positively contribute to the improving the engagement of the user. In addition to this, a greater understanding of how to implement these new concepts has been obtained, which allows them to be tested more thoroughly in future to establish their effectiveness and reduce the overhead of the systems to obtain more reliable results.

5.1   Future Work

As mentioned in the previous section, several areas have been identified where the implementation of these ideas could be improved, such as reducing the amount of time a user has to wait to see the next generation. With the information gained from this pilot study, improved implementations of these systems can be created, which can then be used to run a full-scale experiment to get a clearer understanding of the benefits of the introduced concepts.

In addition to further application of these systems to reducing user fatigue in evolutionary art, they offer an exciting base from which non-related research can be completed. Various aspects of these systems cover a wide range of topics within computer science, such as immersion and presence, 3D drawing methods or the investigation of 3D aesthetics.

References

1. Henshilwood, C.S., D'Errico, F., van Niekerk, K.L., Dayet, L., Queff-elec, A., Pollarolo, L.: An abstract drawing from the 73,000-year-old levels at Blombos Cave, South Africa. Nature **562**(7725), 115–118 (2018). http://www.nature.com/articles/s41586-018-0514-3
2. Ekárt, A., Sharma, D., Chalakov, S.: Modelling human preference in evolutionary art. In: Di Chio, C., et al. (eds.) EvoApplications 2011, Part II. LNCS, vol. 6625, pp. 303–312. Springer, Heidelberg (2011). https://doi.org/10.1007/978-3-642-20520-0_31
3. Takagi, H.: Interactive evolutionary computation: fusion of the capabilities of EC optimization and human evaluation. Proc. IEEE **89**, 1275–1296 (2001)
4. Lourenço, N., Assunção, F., Maçãs, C., Machado, P.: EvoFashion: customising fashion through evolution. In: Correia, J., Ciesielski, V., Liapis, A. (eds.) EvoMUSART 2017. LNCS, vol. 10198, pp. 176–189. Springer, Cham (2017). https://doi.org/10.1007/978-3-319-55750-2_12
5. Cohen, M.W., Cherchiglia, L., Costa, R.: Evolving Mondrian-Style artworks. In: Correia, J., Ciesielski, V., Liapis, A. (eds.) EvoMUSART 2017. LNCS, vol. 10198, pp. 338–353. Springer, Cham (2017). https://doi.org/10.1007/978-3-319-55750-2_23
6. Secretan, J., Beato, N., D'Ambrosio, D.B., Rodriguez, A., Campbell, A., Stanley, K.O.: Picbreeder: evolving pictures collaboratively online. In: Proceedings of the SIGCHI Conference on Human Factors in Computing Systems, pp. 1759–1768. ACM (2008)
7. O'Brien, H.L., Toms, E.G.: What is user engagement? A conceptual framework for defining user engagement with technology. J. Am. Soc. Inf. Sci. Technol. **59**(6), 938–955 (2008)
8. Davies, E., Tew, P., Glowacki, D., Smith, J., Mitchell, T.: Evolving atomic aesthetics and dynamics. In: Johnson, C., Ciesielski, V., Correia, J., Machado, P. (eds.) EvoMUSART 2016. LNCS, vol. 9596, pp. 17–30. Springer, Cham (2016). https://doi.org/10.1007/978-3-319-31008-4_2
9. Sureephong, P., Puritat, K., Chernbumroong, S.: Enhancing user performance and engagement through gamification: case study of aqua republica. In: 2016 10th International Conference on Software, Knowledge, Information Management & Applications (SKIMA), pp. 220–224. IEEE (2016)
10. Essential Facts About the Computer and Video Game Industry - The Entertainment Software Association. http://www.theesa.com/about-esa/essential-facts-computer-video-game-industry/
11. Hastings, E.J., Guha, R., Stanley, K.: Evolving content in the galactic arms race video game. In: Computational Intelligence and Games, pp. 241–248. IEEE (2009)
12. Risi, S., Lehman, J., D'Ambrosio, D., Hall, R., Stanley, K.: Combining search-based procedural content generation and social gaming in the Petalz video game. In: Aiide. AAAI (2012)
13. Easton, E.: Investigating User Fatigue in Evolutionary Art. Masters, Aston University (2018)
14. VIVE — Buy VIVE Hardware. https://www.vive.com/eu/product/
15. SteamSpy - All the data and stats about Steam games. https://steamspy.com/
16. Galván-López, E., McDermott, J., O'Neill, M., Brabazon, A.: Defining locality as a problem difficulty measure in genetic programming. Genet. Program. Evolvable Mach. **12**(4), 365–401 (2011)

17. Re, A., Castelli, M., Vanneschi, L.: A comparison between representations for evolving images. In: Johnson, C., Ciesielski, V., Correia, J., Machado, P. (eds.) Evo-MUSART 2016. LNCS, vol. 9596, pp. 163–185. Springer, Cham (2016). https://doi.org/10.1007/978-3-319-31008-4_12

18. Colton, S., Cook, M., Raad, A.: Ludic considerations of tablet-based evo-art. In: Di Chio, C., et al. (eds.) EvoApplications 2011, Part II. LNCS, vol. 6625, pp. 223–233. Springer, Heidelberg (2011). https://doi.org/10.1007/978-3-642-20520-0_23

19. Hull, M., Colton, S.: Towards a general framework for program generation in creative domains. In: Programme Committee and Reviewers, p. 137 (2007)

20. Tilt Brush by Google. https://www.tiltbrush.com/

21. Easton, E.: C# Tilt Brush Toolkit. https://github.com/Prystopia/c-sharp-tiltbrush-toolkit

EvoChef: Show Me What to Cook!
Artificial Evolution of Culinary Arts

Hajira Jabeen[1(✉)], Nargis Tahara[1], and Jens Lehmann[1,2]

[1] Informatik III, University of Bonn, Bonn, Germany
jabeen@cs.uni-bonn.de, nargistahara@gmail.com
[2] Fraunhofer IAIS, Sankt Augustin, Germany
jens.lehmann@iais.fraunhofer.de

Abstract. Computational Intelligence (CI) has proven its artistry in creation of music, graphics, and drawings. EvoChef demonstrates the creativity of CI in artificial evolution of culinary arts. EvoChef takes input from well-rated recipes of different cuisines and evolves new recipes by recombining the instructions, spices, and ingredients. Each recipe is represented as a property graph containing ingredients, their status, spices, and cooking instructions. These recipes are evolved using recombination and mutation operators. The expert opinion (user ratings) has been used as the fitness function for the evolved recipes. It was observed that the overall fitness of the recipes improved with the number of generations and almost all the resulting recipes were found to be conceptually correct. We also conducted a blind-comparison of the original recipes with the EvoChef recipes and the EvoChef was rated to be more innovative. To the best of our knowledge, EvoChef is the first semi-automated, open source, and valid recipe generator that creates easy to follow, and novel recipes.

Keywords: Recipe · Evolutionary algorithms · Culinary art · Genetic Algorithm

1 Introduction

Culinary art presents itself as an attractive example of a complex art that combines ingredients and spices, blending them into amazing flavours by applying a variety of cooking methods like baking, grilling or frying. It is considered an intricate combination of art and science that handles flavour, texture, nutrients, aroma, and results in a recipe that is edible, healthy and presentable, simultaneously. Over the years, human experts (chefs) have proudly produced wonderful recipes that use novel ingredients, methods, and spices resulting in culinary masterpieces.

Individual cultures have unique preferences for the use of ingredients, spices, cooking methods and their combinations to build a recipe [1]. Different cultures value diverse outlooks on food, such as texture, taste or healthiness, which impacts their cuisine.

© Springer Nature Switzerland AG 2019
A. Ekárt et al. (Eds.): EvoMUSART 2019, LNCS 11453, pp. 156–172, 2019.
https://doi.org/10.1007/978-3-030-16667-0_11

Most of the recipes available on the internet share region or culture-specific cuisines that follow the concept of "good food" of the person who introduces the recipe. Therefore, we miss the unlimited number of possible patterns that can be formed by mixing different cooking techniques, ingredients and spices from different cultures. This innovation remains limited owing to the unlimited number of possible combinations of the ingredients and spices, and availability of these ingredients in different parts of the world.

The role of intelligence in arts or creativity of machines has been a subject of discussion since long. Computational Intelligence (CI) is successfully playing a prominent role in many applications of creative arts. This includes generation of new music [2], compositions [3], art [4] and graphics [5]. A few approaches (e.g. IBM Chef-watson [6,7], Edamam [8], Covercheese [9]) have tried to automatically create or evolve recipes but they have remained limited.

In this paper, we aim to bridge this research-gap by using CI to create, intermix, evolve and optimize the recipes taken from different cuisines. We hypothesize-and-test, "Can Computational Intelligence make it possible to intermix the regional foods from different cultures to create novel works-of-art that are similar yet different from the original recipes?".

We propose "EvoChef", that gathers recipes from multiple cuisines and evolves them to create new recipes with the objective of feasibility and novelty of the resulting recipes. The main contributions of the paper are as follows:

1. An open source software to create and evolve novel recipes
2. Encoding of recipes as trees that can be evolved using genetic algorithm
3. A first proof of concept that combines different regional cosines
4. The resulting recipes:
   (a) Are understandable
   (b) Can be cooked and eaten
   (c) Use common ingredients
5. A system that can be easily extended to evolve novel recipes

To the best of our knowledge, Evochef is the first open source[1], robust and generic culinary artist that produces new recipes that are novel, valid, edible and comprehensive enough to be followed and cooked.

This paper is organized as follows. In Sect. 2, we discuss previous research related to recipe generation. In Sects. 3 and 4, we present the working of EvoChef and results. Section 5 wraps up the discussion, results, and presents future perspectives of this work.

1.1 Preliminaries

Genetic Algorithms. Genetic Algorithms (GA) [10] are heuristics built upon Darwin's idea of natural selection and survival of the fittest for evolution. GAs mimic evolving a population of random solutions to an optimization problem. These solutions, having abstract representations (called chromosomes or the

---

[1] https://github.com/SmartDataAnalytics/EvoChef.

genotype or the genome) of candidate solutions (called individuals, creatures, or phenotypes), are evolved for a certain number of generations by exploiting the information from the fitness of all individuals and exploring the search space using recombination operators. The chromosome or genotype of the solution must be carefully designed to represent a viable possible solution, that can undergo recombination operators (mutation or crossover) and has ability to explore the search space for the optimal solution. We briefly cover a few important concepts of GA below:

1. **Initial Population**: The initial population is created randomly from the solution space.
2. **Fitness**: This is the objective function to optimize. It determines the quality of an individual and helps in estimating the probability of being selected as a parent.
3. **Selection**: The process used to select the two (relatively fit) parents for recombination. The selection mechanism mostly takes the fitness of individuals into account.
4. **Crossover**: This operator combines (swaps) chromosomes of two parent solutions to create two new offspring solutions
5. **Mutation**: This operator injects new traits into a chromosome by randomly updating a portion of the selected chromosome.
6. **Termination**: The criteria to stop the evolution process e.g. the number of generations(iterations), or desired (average/maximum) fitness.

The pseudo-code of GA is detailed in Algorithm 1.

Algorithm 1. Genetic Algorithms

Result: Best solution
1 Optimization problem
2 *START*
3 *Generate the initial population*
4 *Compute fitness of the population*
5 **while** *termination condition* **do**
6 | *Selection*
7 | *Crossover*
8 | *Mutation*
9 | *Compute Fitness*
10 *STOP*

We have chosen to work with GA in this work, as it has been proven efficient [11] in searching for complex search spaces while maintaining the population diversity.

GraphX. GraphX [12,13] is a graph processing library of Apache Spark [14] for graph parallel computation. GraphX supports graph abstraction of a directed graph with properties attached to its vertices and edges. Each vertex in the graph is identified by a unique identifier called VertexId. The edges are represented by the identifiers of their source and destination vertex. GraphX provides a generic interface to work with properts graphs. We have used property graph to represent recipes as trees. The nodes in this graph represent steps and properties represent the type, names, quantities or status. The fact that this additional information can be a part of the node has made easier to perform recombination operators. We detail this information in the later sections when we discuss the recombination operators.

2   Related Work

The use of CI in culinary arts has remained limited. This is partially due to the complexity and multimodality inherent in recipe generation. There have been a few attempts to develop computer-generated culinary arts but they have remained far from being optimal. One of the most prominent efforts was by IBM [15], named "IBM Chef-Watson" [6] presented in 2014. IBM built a system that produced novel recipes by introducing new ingredients in existing recipes from Bon Appetit [29], and other resources. Their constrained system provided an interface, where the user could select the dish, the cuisine type, and the ingredients. Based on these choices, the system showed new recipes to the users. The quality of the recipe was determined based on its novelty and aroma. The novelty was measured as the deviation from the common recipes, and the aroma was measured by evaluating the chemical properties of flavour molecules used in the recipe [16]. Despite the imagined level of effort involved in IBM-research, the Chef-Watson was unable to produce valid recipes; The instructions were incomplete or incomprehensible and the quantities of the ingredients were missing. The recipes were taken from bon-appetite, and therefore, were mainly western cuisine recipes. It focused novelty using chemical combinations of the ingredients and therefore, often resulted in hard to find ingredients. Overall IBM Chef Watson lacked a few key elements for regular cooking. The technical details of the working of Chef Watson were not formally published. One of the user reviews states "the combination of ingredients suggested by Watson was so unfamiliar and there were many mistakes in the recipes[2]. This project seems to be canned now, and the web page[3] is no longer available.

Evolutionary Meal Management Algorithm (EMMA) [9] is another attempt to generate the recipes using machine learning algorithms. EMMA can automatically generate recipes for new food items and improve them based on user's feedback about their taste. In the early versions of EMMA, the system was unable to detect edible ingredients and produced inedible recipes e.g. *"1 cup skillet, chop skillet and serve"*. The later generations, look a little better but they are very basic and often miss clear instructions or quantities. Moreover, the work seems experimental and there are no published papers about the underlying approach.

---

[2] https://daljiblog.wordpress.com/2016/11/30/how-chef-watson-calculates-deliciousness/.

[3] http://www.ibmchefwatson.com.

Erol et al. [17] have developed an approach to discover novel ingredient combinations for the salads. The system is designed in two steps: First, a statistical model is constructed to rank recipes using a deterministic classifier that, given two recipes, predicts which one is better. Second, they experimented with various search algorithms to explore the salad recipe space and discover novel ingredient combinations. Like Chef-Watson; their focus is also on novelty. In the end, the algorithm only suggests ingredients and ignores any instructions or the quantities of ingredients, which is one of the crucial elements of any recipe.

Most of the above-mentioned approaches do not always present valid recipes to their users and focus mostly on novelty by providing good food pairing. They do not emphasise easy to follow recipes, easy ingredients, or regional cuisines. On the contrary, our work attempts to cover novelty, cultures, usability, and comprehensiveness of resulting recipes.

Table 1. Components of a recipe

Component	Description
Main ingredient	The main ingredient is a major ingredient of the recipe. e.g. in all types of rice dishes, rice is the main ingredient
Side ingredient	Side ingredients are the remaining ingredients of the recipe minus the main ingredient
Spices	A spice is a type of side ingredient used to add flavour to the ingredients in the recipe
Steps	These are the cooking instructions
Pre-processing	A step required before actual cooking. For example, to fry onions, we first need to peel and cut them into chunks
Post-processing	The post-processing steps mix the individual ingredients together and do some processing e.g. bake, fry or saute

3 Proposed Methodology

3.1 Preliminaries

Components of a Recipe. We have represented the recipes such that each recipe contains a main ingredient, some side ingredients, spices, and instructions for cooking that are named as steps. A recipe can also have some pre-processing or post-processing steps. Pre-processing steps are performed on ingredients before mixing and post-processing is performed after mixing the ingredients. Table 1 explains the recipe structure in more detail.

Data Collection and Cleaning.
In order to achieve diversity in our
initial population, we have collected
our recipes from different cooking
websites including Yummly.com [18],
Allrecipes.com [19], Recipes-plus.com
[20], Geniuskitchen.com [21], Simplyrec
ipes.com [22], Omnivorescookbook.com
[23] and Greenevi.com [24]. Our recipe
data consists of Southern, American,
Italian, Spanish, Hungarian and Chinese recipes.

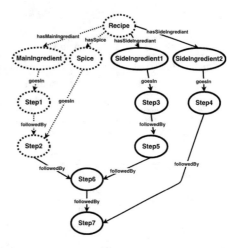

Fig. 1. Graph representation of an arbitrary recipe.

Due to the fact that the recipes are
written by different users of these popular web pages, they do not possess
a regular format or consistent structure. e.g. the pre-processing steps for
the ingredients are either written in the
instructions or mentioned besides the ingredient name. Therefore, we have
cleaned the data to make it feasible with our proposed recipe structure. The
following steps are performed on each recipe to make it ready to be ingested by
EvoChef:

1. Multiple quantity options were converted to a single quantity. For example,
   if there is "3–4 potatoes" in the ingredient list, it is converted to either "3
   potatoes" or "4 potatoes".
2. Properties of ingredients were extracted and assigned to each ingredient
   accordingly. Properties include ingredient name, quantity, measurement unit,
   ingredient type (main ingredient/side ingredient) and usedIn (whether the
   ingredient is used in the main-process or side-process).
3. If pre-processing steps are along with ingredients, they were separated and
   added to the instructions. For example, if "4 potatoes (cut into small chunks)"
   is in the ingredient list. We have removed the "cut into small chunks" from
   the ingredient list and added this as another step in the instructions.
4. Each step in the instruction is assigned the property usedIn (whether the step
   is part of the main process or side process).
5. The root node, recipe, has the properties: recipeName, the totalTime required
   to cook the recipe, numberOfServings, category of the recipe (Main dish/Side
   dish) and the state of the main ingredient after the main process has been
   applied (e.g. raw/half cooked/cooked).

Considering a fully automated, NLP based, approach beyond the scope of
this work, we have adopted a semi-automated approach to curate and clean the
recipe data.

Table 2. Edges/Relations properties

Properties	Description
hasSideIngredient	Domain: Recipe
	Range: Ingredient
	Description: The recipe has a side ingredient
hasMainIngredient	Domain: Recipe
	Range: Ingredient
	Description: The recipe has a main ingredient
hasSpice	Domain: Recipe
	Range: Ingredient
	Description: Recipe has a spice
followedBy	Domain: step, post-step
	Range: step, post-step, last-step
	Description: A step is followed by another step
goesIn	Domain: main Ingredient, side ingredient, spice
	Range: Step
	Description: The ingredient or spice goes in (added to) a step

3.2   Evolution of the Recipes

This section describes the overall evolution process and representation decision taken in this work.

Solution Encoding. A recipe contains a set of ingredients, spices and cooking steps. We have used the property graph of GraphX to encode recipes, where vertices in the graph represent important components of the recipes (e.g. ingredients, spices and method) that make up the textual representation of a recipe, and edges define relations (e.g. hasSpice, goesIn etc.) between them. This representation corresponds to the genotype encoding to ensure credible evolution. The cooking process of a recipe can be divided into two sub-processes; a main process and a side process. The main process contains the cooking steps of the main ingredient that are required prior to mixing with the rest of the ingredients. The side process represents the pre-processing and cooking steps of the side ingredients that are later to be mixed with the main process. In Fig. 1, an example recipe is shown with one main ingredient, two side ingredients and one spice. The nodes and edges represented with the dotted line below the root show the main process. The right hand side of the root with solid line represents the side ingredients and the process. Tables 3 and 2 describe the properties of vertices and edges of the graph respectively.

Initial Population. The initial population is generated by randomly intermixing (crossover) the recipes taken from the recipe pages. This was done to avoid the first generation to have the best fitness, as we selected the best rated recipes.

Fitness Evaluation. The challenging part of our work is the evaluation of the fitness of a recipe. Human experts can best judge a recipe by either tasting or even examining the description of the recipe. We listed the recipes generated by our system on a dedicated website[4] and gathered user ratings (friends, colleagues, and research partners) from a range of users as 1–5 stars. This fitness-rating is used as the fitness function by EvoChef.

Selection. We have used tournament selection as the parent selection mechanism. To produce a child, we run the tournament two times for each parent. Each parent goes through a compatibility check before performing the recombination operator. The compatibility check tests if "cookedType_mainIngredient" property of both parent recipes is cooked, to ensure that the children produced are valid and we are not mixing cooked ingredients with uncooked ones, this might result in an uncooked ingredient.

Crossover. We have used the fixed point crossover operator in EvoChef. The main and side processes of the parents are swapped to create two children with different valid recipes. Table 4 shows two parent recipes.

Parent recipe#01 has one main ingredient and four side ingredients. The main process of the recipe is represented in italic text while rest of the text shows the side process of the recipe. Figure 2 shows the graph representation of the first recipe. The edge between nodes and steps show the ingredients of the step. The Parent recipe#02 is represented as a graph in Fig. 3. Now we have two parents recipes (phenotypes) encoded as graphs (genotypes). We can apply crossover and mutation on these graphs to generate new off springs. Table 5 shows the children phenotypes after recombination of the parent recipes from Table 4. Figure 4 shows the genotypes of child_01_02. Figure 5 is graph representation of second child child_02_01. Out of the two children, the fittest, is returned to be added to the following generation.

Generation of Recipe Names. The child recipe name is generated from its parents. If there is any side ingredient name from one parent recipe, it is replaced with the name of the side ingredients from the second parent. Following this, in child_01_02, recipe name from recipe#01 will be taken and if there is the name of any side ingredient from recipe#01 in the name of recipe; it will be replaced with the side ingredient of recipe#02, as in new recipe main ingredient of recipes#01 and side ingredients if recipe#02 are being used.

---

[4] http://www.machinegeneratedrecipes.de.

Name of recipe#01: Tibetan potato curry
Name of recipe#02: Glazed sweet potatoes with brown sugar
Name of child_01_02: Tibetan potato curry with brown sugar
Name of child_02_01: Glazed potatoes

Listing 1.1. Child name generation

Table 3. Vertex properties

	Recipe properties
Property	Description
RecipeName	Name of the recipe
cookTime	Time required to cook recipe
Servings	To how many people the recipe serves
Category	Category of the recipe. For example, main dish, side dish, dessert etc.
cookedType_mainIngredient	Whether the main ingredient is raw, half cooked or cooked after the main process is finished. In other words, it is the condition of the main ingredient before it is going to be added with other side ingredients
	Ingredient properties
Property	Description
IngredientName	Name of the ingredient
Quantity	Quantity of the ingredient
measurementUnit	Measurement Unit of the ingredient, for example, kg, cups, tablespoon etc.
ingredientType	Main ingredient, side ingredient or spice
usedIn	Whether the ingredient is the part of main process or side process
	Steps properties
Property	Description
Description	Cooking description of the step, The step can also be a post-step or last-step
usedIn	Whether the step is a part of the main process or side process

These new names remain compatible with the content of the new recipe. If there are no side ingredients in the first parent, the side ingredients from the second recipe are added at the end of the recipe name with propositions "with" or "and". An example is shown in Listing 1.1.

Name Generalization. At the end of the process of offspring generation, the name of the main ingredient in the cooking instructions of the recipe is generalized to avoid confusion in the new recipe. For example, in our case, different kinds of potatoes are used in different recipes including Red Potatoes, Russet Potatoes, sweet potatoes, etc. As they are added in another recipe that might be referring to the potatoes with another name. To avoid confusion, all type of potatoes is replaced with "potatoes" in the cooking instructions. As shown in Table 5, in step 4 of Child_02_01, sweet potato is replaced with the potato.

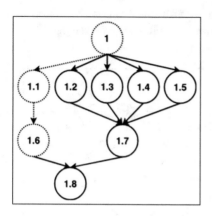

Fig. 2. Graph representation of parent recipe#01

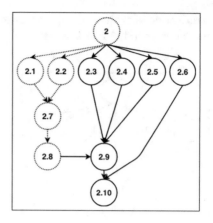

Fig. 3. Graph representation of parent recipe#02

Table 4. Parent recipes

Parent recipe#01	Parent recipe#02
Glazed Sweet Potatoes with Brown Sugar [25]	**Mashed Red Potatoes With Garlic And Parmesan** [26]
Ingredients:	**Ingredients:**
brown sugar, water, butter, salt, *sweet potatoes*	*red potatoes, garlic cloves,* butter, milk, salt, parmesan cheese
Instruction:	**Instruction:**
1. *Peel the sweet potatoes and cut them into 0.5 in. to 1-in. thick slices. Place the sweet potato slices in a saucepan and cover with water. Bring to a boil and cook for about 12 min, or until just tender.*	1. *Put potatoes and garlic in a large pan. Cover with water. Bring to a boil.* 2. Reduce heat and simmer for 25 min, until potatoes are tender. Drain well 3. Mash with the butter, milk, and salt 4. Stir in the parmesan cheese
2. In a heavy skillet, combine brown sugar, water, butter, and salt. Simmer over low heat for 5 min	
3. Add the sliced sweet potatoes to the brown sugar mixture. Simmer for 10 min, or until well glazed, turning frequently to keep them from scorching	

Mutation. In order to introduce diversity through mutation, we have developed an ingredient replacement table shown in Table 6. This is a simplified version of the network provided by food-network [27]. An ingredient of the selected recipe is picked randomly and replaced by its substitution from the table. This type of mutation does not make any changes in the cooking method itself, but is still able to introduce novelty through replacing the ingredients. Table 7 represents an example where butter in the recipe is replaced with margarine in the given recipe.

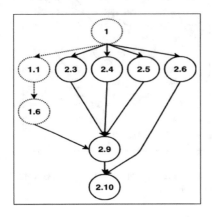

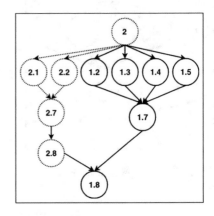

Fig. 4. Graph representation of Child_01_02

Fig. 5. Graph representation of Child_02_01

Table 5. Child recipes

Child recipe_01_02	Child recipe_02_01
Glazed Sweet Potatoes with parmesan cheese	**Mashed Red Potatoes With brown sugar And butter**
Ingredients:	**Ingredients:**
butter, milk, salt, parmesan cheese, *sweet potatoes*	*red potatoes, garlic cloves,* brown sugar, water, butter, salt
Instruction:	**Instruction:**
1. *Peel the sweet potatoes and cut them into 0.5 in. to 1-in. thick slices. Place the sweet potato slices in a saucepan and cover with water. Bring to a boil and cook for about 12 min, or until just tender.* 2. Mash with the butter, milk, and salt 3. Stir in the parmesan cheese	1. *Put potatoes and garlic in lg pan. Cover with water. Bring to a boil.* 2. *Reduce heat and simmer for 25 min, until potatoes are tender. Drain well.* 3. In a heavy skillet, combine brown sugar, water, butter, and salt. Simmer over low heat for 5 min 4. Add the sliced potatoes to the brown sugar mixture. Simmer for 10 min, or until well glazed, turning frequently to keep them from scorching

Table 6. Sample ingredient substitute table

Ingredient	Substitute
Arrowroot starch	Flour, cornstarch
Baking mix	Pancake mix, Biscuit Mixture
Beer	Nonalcoholic beer, chicken broth
Bread crumbs	Cracker crumbs, matzo meal, ground oats
Butter	Margarine, shortening, vegetable oil, lard
Buttermilk	Yogurt
Cheddar cheese	Shredded Colby cheddar, shredded Monterey Jack cheese
Chervil	Chopped fresh parsley
Chicken base	Chicken broth, chicken stock
Cocoa	Unsweetened chocolate
Cottage cheese	Farmer's cheese, ricotta cheese
Egg	Silken tofu pureed, mayonnaise
Evaporated milk	Light cream
Garlic	Garlic powder, granulated garlic
Honey	Corn syrup, light treacle syrup
Lemon juice	Vinegar, white wine, lime juice
Onion	Green onions, shallots, leek

4   Evaluation

Due to limited available data and dependence of our fitness function on expert input, our experiments have mainly remained limited, but they have yielded promising results nonetheless. The initial population size is eight. with 0.9 crossover rate and 0.1 mutation rate. We evolve the GA until 85% of the population has achieved the fitness rating above 4, with 5 being the highest and 0 being the lowest fitness-rating.

4.1   Generated Recipes

While we have presented some child recipes in the earlier sections. All the generated recipes are listed on our webpage[5]. We have gathered fitness-ratings and comparison scores for our recipes from the users. These users were invited over social media channels, friends, colleagues and research partners.

---

[5] www.machinegeneratedrecipes.de.

Table 7. Mutation by using ingredient substitution

Original recipe	Recipe after mutation
Glazed Sweet Potatoes with parmesan cheese	**Glazed Sweet Potatoes with parmesan cheese**
Ingredients: ~~butter~~, milk, salt, parmesan cheese, sweet potatoes	**Ingredients:** *margarine*, milk, salt, parmesan cheese, sweet potatoes
Instruction: 1. Peel the sweet potatoes and cut them into 0.5 in. to 1-in. thick slices. Place the sweet potato slices in a saucepan and cover with water. Bring to a boil and cook for about 12 min, or until just tender 2. Mash with the ~~butter~~, milk, and salt 3. Stir in the parmesan cheese	**Instruction:** 1. Peel the sweet potatoes and cut them into 0.5 in. to 1-in. thick slices. Place the sweet potato slices in a saucepan and cover with water. Bring to a boil and cook for about 12 min, or until just tender 2. Mash with the *margarine*, milk, and salt 3. Stir in the parmesan cheese

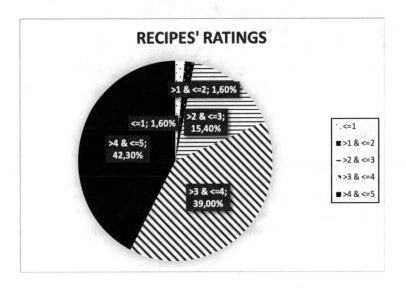

Fig. 6. Recipe ratings in the initial generation

4.2   Recipe Ratings (Fitness)

This section overviews the fitness-ratings of the evolving recipes. Figure 6 shows the overall fitness of the initial population. The fitness of the first three generations is shown in Fig. 7 and it can be observed that the average fitness of the population successfully improved with the number of generations.

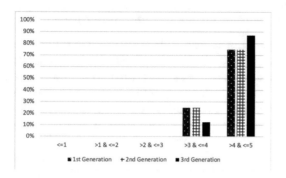

Fig. 7. Child population ratings in 1st, 2nd and 3rd generation

Evochef produced a total of 123 recipes from all the possible combinations of recipes in the initial population, from which more than 80% recipes have good rating (more than 3 out of 5) and 42% recipes have an excellent rating (more than 4 out of 5). 7% of the produced child were rejected by compatibility function hence dropped.

4.3 Novelty

To compare the novelty in our newly generated recipes, we made a blind comparison of recipes. We listed a randomly selected original recipe with an EvoChef recipe, and users were asked to select a recipe with more novelty. Three of 12 comparisons rated original recipe as more novel while 8 of the new recipes received more votes as a novel recipe. One of the recipes received equal votes for original and new recipes (Table 8).

Table 8. Novelty of EvoChef recipes

Original	EvoChef	Both are equal	I do not know
36.9%	59.2%	1.90%	2.00%

4.4 Comparison with Parent Recipes

All the recipes selected as the input to EvoChef had the rating of '5' on their parent webpages. In order to have a fair comparison, we copied the same recipes on our page and also gathered ratings from the same users. At the end of our experiments, we found that the average ratings of parent recipes is 4 and the average rating of EvoChef recipes in the last generation is 4.5.

4.5   Comparison with Other Recipe Generators

IBM Chef-Watson is no longer active. However, we found some reviews [28] about the recipes that "Calls for wasabi powder (never used), **shelled green peas** (212 cup shelled green peas) **cut into 3/4" pieces**. Then placed on a barbecue." Cover cheese [9] has recipes like "Ingredients med okra, lot sugar and Instructions: **boil: sugar okra sugar**" Or "Ingredients: Small angel_food, Small eggplant and Instructions: 1. Slice:angel_food, 2. eggplant angel_food" Both of these examples do not create a valid recipe. Erol et al. [17], only targets salads and their recipes look like **"Cherry, chive, granny smith apple, mushroom, onion powder, pine nut, salsa, salt"** Their salad recipes have no instructions or the quantities of the ingredients.

5   Conclusion

Our work has investigated the possibility of using evolutionary algorithms in culinary arts to develop a system that produces valid and novel recipes. The encoding of the recipes as graphs has provided an optimal representation of recipes to undergo genetic operators like crossover and mutation. The fact that our initial recipes are highly rated (5) and are taken from popular webpages. They mostly use easy-to-find ingredients. This important feature is also reflected in the recipes generated by Evochef.

The results produced by EvoChef are complete and precise. While we have constrained ourselves to focus only potato recipes. This approach can be extended to different kinds of recipes by automatically extracting recipe-data using natural language processing approaches.

Our limited data and the bottleneck of human feedback for the evaluation forced us to stop the evolution process earlier. Machine learning approaches can be employed to predict the ratings of the child recipes automatically. Currently, we ignore some relevant features like flavor information for ingredient-pairing, nutritional information, or the texture of the recipe. Extending the approach with this information could yield healthier food. In conclusion, our preliminary work has produced interesting results in the this under-represented area or evolving culinary arts, and there is a range of possibilities to extend this work.

Acknowledgement. This work is partly supported by the EU Horizon2020 projects BigDataOcean (GA no. 732310), LAMBDA (GA no. 809965) and Boost4.0 (GA no. 780732).

References

1. Kim, K., Chung, C.: Tell me what you eat, and i will tell you where you come from: a data science approach for global recipe data on the web. IEEE Access **4**, 8199–8211 (2016)
2. De Prisco, R., Zaccagnino, R.: An evolutionary music composer algorithm for bass harmonization. In: Giacobini, M., et al. (eds.) EvoWorkshops 2009. LNCS, vol. 5484, pp. 567–572. Springer, Heidelberg (2009). https://doi.org/10.1007/978-3-642-01129-0_63
3. Scirea, M., Togelius, J., Eklund, P., Risi, S.: Affective evolutionary music composition with metacompose. Genet. Program. Evolvable Mach. **18**, 1–33 (2017)
4. Misztal, J., Indurkhya, B.: A computational approach to re-interpretation: generation of emphatic poems inspired by internet blogs (2014)
5. Lewis, M.: Evolutionary visual art and design. In: Romero, J., Machado, P. (eds.) The Art of Artificial Evolution, pp. 3–37. Springer, Heidelberg (2008). https://doi.org/10.1007/978-3-540-72877-1_1
6. IBM Chef Watson. http://www.ibmchefwatson.com
7. Pinel, F.: What's cooking with chef watson? An interview with lav varshney and james briscione. IEEE Pervasive Comput. **14**(4), 58–62 (2015)
8. EDAMAM. https://www.edamam.com/
9. Cover:Cheese. https://covercheese.appspot.com/
10. Mitchell, M.: An Introduction to Genetic Algorithms. MIT Press, Cambridge (1996)
11. Anderson-Cook, C.M.: Practical Genetic Algorithms (2005)
12. Xin, R.S., Gonzalez, J.E., Franklin, M.J., Stoica, I.: GraphX: a resilient distributed graph system on spark. In: First International Workshop on Graph Data Management Experiences and Systems, GRADES 2013, pp. 2:1–2:6. ACM, New York (2013)
13. GraphX Programming Guide. https://spark.apache.org/docs/latest/graphx-programming-guide.html#connectedcomponents
14. Zaharia, M., et al.: Apache Spark: a unified engine for big data processing. Commun. ACM **59**(11), 56–65 (2016)
15. IBM. https://www.ibm.com
16. Bhatia, A.: A new kind of food science: how IBM is using big data to invent creative recipes (2013). https://www.wired.com/2013/11/a-new-kind-of-food-science/. Accessed 01 Mar 2018
17. Cromwell, E., Galeota-Sprung, J., Ramanujan, R.: Computational creativity in the culinary arts (2015)
18. Yummly. https://www.yummly.com/
19. Allrecipes. https://www.allrecipes.com
20. RecipesPlus. http://recipes-plus.com/
21. Genius Kitchen - Recipes, Food Ideas And Videos. https://www.geniuskitchen.com
22. Simply Recipes. https://www.simplyrecipes.com/
23. Omnivore's Cookbook. https://omnivorescookbook.com/
24. Green Evi. http://greenevi.com
25. Rattray, D.: Glazed sweet potatoes with brown sugar. https://www.thespruceeats.com/glazed-sweet-potatoes-with-brown-sugar-3061580?utm_campaign=yummly&utm_medium=yummly&utm_source=yummly. Accessed Feb 2018

26. MizzNezz: Mashed red potatoes with garlic and parmesan. http://www.geniuskitchen.com/recipe/mashed-red-potatoes-with-garlic-and-parmesan-34382#activity-feed. Accessed Feb 2018

27. Common Ingredient Substitutions (Infographic). http://dish.allrecipes.com/common-ingredient-substitutions/

28. IBM-ICE. https://www.reddit.com/r/IAmA/comments/3id842/we_are_the_ibm_chef_watson_team_along_with_our/

29. Bon Appetit. https://www.bonappetit.com/

Comparing Models for Harmony Prediction in an Interactive Audio Looper

Benedikte Wallace$^{(\boxtimes)}$ and Charles P. Martin

RITMO Centre for Interdisciplinary Studies in Rhythm, Time, and Motion,
Department of Informatics, University of Oslo, Oslo, Norway
{benediwa,charlepm}@ifi.uio.no

Abstract. Musicians often use tools such as loop-pedals and multitrack recorders to assist in improvisation and songwriting, but these tools generally don't proactively contribute aspects of the musical performance. In this work, we introduce an interactive audio looper that predicts a loop's harmony, and constructs an accompaniment automatically using concatenative synthesis. The system uses a machine learning (ML) model for harmony prediction, that is, it generates a sequence of chords symbols for a given melody. We analyse the performance of two potential ML models for this task: a hidden Markov model (HMM) and a recurrent neural network (RNN) with bidirectional long short-term memory (BLSTM) cells. Our findings show that the RNN approach provides more accurate predictions and is more robust with respect to changes in the training data. We consider the impact of each model's predictions in live performance and ask: "What is an accurate chord prediction anyway?"

Keywords: RNN · Deep learning · Music interaction ·
Machine improvisation

1 Introduction

Though theoretically, there are no "wrong" chord choices, most people would agree that, for a given melody, some chord choices create a dissonance that does not sound particularly good. In order to decide precisely what chords should accompany a melody, human musicians apply knowledge from practical experience as well as formal rules of music theory. Stylistically appropriate chord choices rely on relationships between the melodic notes and chord, as well as temporal dependencies between chords in a sequence: what comes before and what comes next are both important. In this work[1] we apply machine learning (ML) to this problem in order to create a harmony prediction module for an interactive audio looping system. This predictive songwriting with concatenative accompaniment (PSCA) system uses a loop-pedal style interface to allow a musician to record a vocal melody, and then constructs an accompaniment layer of audio from their own recorded voice (Fig. 1).

---

[1] This research is an extension of the author's master's thesis [24].

© Springer Nature Switzerland AG 2019
A. Ekárt et al. (Eds.): EvoMUSART 2019, LNCS 11453, pp. 173–187, 2019.
https://doi.org/10.1007/978-3-030-16667-0_12

Fig. 1. Predictive Songwriting with Concatenative Accompaniment system setup: laptop running the PSCA software, sound card, microphone, headset and Arduino foot switch controller.

We consider two different models for harmony prediction in this research: a hidden Markov model (HMM) [18], the more traditional option, and a bidirectional recurrent neural network [21] with long short-term memory cells [10] (BLSTM), a newer deep neural network model. We analyse these two models from the perspective of predictive accuracy, and find that the BLSTM provides more accurate predictions overall, and is more robust to changes in the training data. Moreover, the BLSTM model provides more "creative" sounding chord sequences which can be further adjusted with a sampling temperature parameter. This makes the BLSTM network much more suitable for our interactive music tool. We discuss the implications of these results, particularly as they relate to the "accuracy" of a chord progression and suggest future directions for training musical machine learning systems to value creative choices.

1.1   The PSCA Looper

The underlying system developed for the PSCA is an audio looper written in Python and controlled by a pair of Arduino-interfaced foot switches. When the program is running the user sings into a microphone and controls recording and playback using the foot switches. The system allows the user to record, loop and play back layers of audio. In order to construct additional harmonies predicted by the models, the recorded audio is segmented, analyzed and added to a note bank according to its pitch. By concatenating and layering the necessary audio segments from the note bank the system is capable of adding new harmonies to the playback loop, creating an accompaniment which changes when the user records new layers. An overview of this process is shown in Fig. 2.

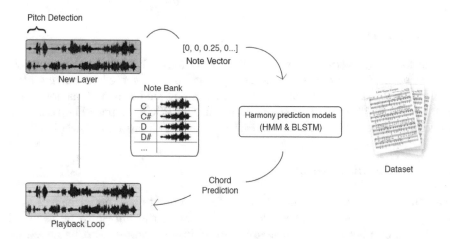

Fig. 2. PSCA system overview: Audio recorded by the user is added to the playback loop together with a chord selected by the harmony prediction models. The harmony is constructed using concatenated segments of the recorded audio.

1.2 Harmony Prediction

At the core of the PSCA system is a module for predicting suitable harmonies for a given sequence of melody notes. This module could be said to engage in *harmony prediction*, as shown in Fig. 3. In this work, we consider the task of predicting suitable harmonies as a sequence-to-sequence prediction problem; given an input sequence of melodic notes, the model must choose a parallel sequence of harmonically appropriate chords. As shown in Fig. 3 we restrict this problem to predicting just one chord for each bar of melody. A human would solve this problem by choosing a sequence of chords that works for the melody, as well as forming a sensible harmonic sequence. Our models will learn how to solve this problem through examples of chord and melody sequences from a data set of lead sheets. How well they perform in the harmony prediction task is evaluated using the accuracy of the predictions against the true values in the data set. Paired with qualitative analysis, the predictive accuracy gives us an idea of how good the models are.

Fig. 3. The harmony prediction problem involves finding suitable chords $(q_1, \ldots, q_6)$ for a given melodic sequence. We consider the limited problem of finding one chord per bar of music.

1.3   Paper Overview

In the following section we will outline previous work on generating musical accompaniments using machine learning. Section 3 is devoted to our methods and materials, descriptions of the data set and our models as well as how these are used to facilitate harmony prediction in the PSCA looper. The results of model evaluation are shown in Sect. 4.1 and further discussed in Sect. 5.

2   Intelligent Loopers

Many musicians use looper pedals and effects to create solo performances with multiple layers of sound. These devices allow an initial phrase to be recorded which is then played back repeatedly and additional phrases can be added on top to build up a complex backing sound. While these devices are very popular, they can only play sounds that have been recorded. More intelligent looping devices have been proposed to overcome this limitation such as the "Reflexive Looper" [17] that modifies the looped phrases to fit a predetermined chord sequence. The Reflexive Looper determines the style of the musician's playing so that the generated accompaniment follows the musician's performance and can distinguish between several playing modes, filling the roles of drummer, bass player or keyboardist as needed.

SongSmith [22] is a system that automatically chooses chords to accompany vocal melodies. By singing into a microphone the user can experiment with different music styles and chord patterns through a user interface designed to be intuitive also for non-musicians. The system is based on a HMM trained on a music database consisting of 298 lead sheets, each consisting of a melody and an associated chord sequence.

JamBot [2] used two LSTM-RNNs. One predicts chord progressions based on a chord embedding similar to the natural language approach of word embeddings used in Google's word2vec [16], the other generates polyphonic music for the given chord progression. The results exhibit long term structures similar to what one would hear during a jam session. Martin and Torresen's RoboJam system [15] used a mixture density RNN (MDRNN) to generate additional layers for short touchscreen performances. This system learned to continue musical touchscreen control data, rather than high-level symbolic music.

Broadly, these systems for intelligently extending looped performances have used two ML architectures: HMMs, and RNNs with LSTM cells. HMMs were used in the successful SongSmith system and thus appeared to be a sufficient model for useful harmony prediction. LSTM-RNNs have been used for music generation [6], and were recently compared to HMMs for the specific task of harmony prediction, however they have rarely been used in interactive music applications [14]. Lim et al. [13] compared a HMM, a deep neural net HMM (DNN-HMM) and a bidirectional LSTM-RNN (BLSTM) for harmony prediction. Their findings suggested a large advantage in terms of accuracy for the BLSTM model over SongSmith's HMM. While using a BLSTM comes at a cost of computational power, such a model is still tractable on everyday computational devices.

Given the discrepancy in accounts for the success of HMM vs. BLSTM models, we decided to implement both architectures for our PSCA Looper. This has allowed us to gain insight into the accuracy of the models both through "accuracy" measures, as well as qualitatively through live performance and experimentation.

3   Building Models for Harmony Prediction

In this section we work towards creating two models of harmony prediction for use in the PSCA Looper using the HMM and BLSTM architectures respectively. While previous interactive systems had applied an HMM, the recent improvement of accessibility for RNNs demands that we consider both options.

3.1   Data Set

We used a data set of 1850 lead sheets in music XML format (.mxl) collected online. These contain both a melody and a corresponding sequence of chords. These were sourced from a data set previously shared at wikifonia.org until 2013. Each lead sheet consists of a monophonic melody and chord notation as well as key and time signature. The data set contains western popular music, with examples from genres like pop, rock, RnB, jazz as well as songs from musical theatre and children's songs. There are examples of songs in both major and minor keys. All songs originally in a major key were transposed to C major and songs in a minor key are transposed to A minor. Approximately 70% of the songs are classified as major keys and 30% as minor keys. Songs that contain key changes were removed from the data set.

The number of unique chords in the data set was 151; however, many of these chords featured very few examples. To reduce the number of chord choices, the data set was processed in two different ways. Our first set reduced each chord to one of five triad types—major, minor, suspended, diminished, and augmented—this ignores sevenths, as well as extended, altered, or added tones, leading to 60 possible chord types (5 triads times 12 root notes). The second approach uses only minor or a major triads, resulting in 24 chord types (2 chord types, 12 root notes). This 24-chord data set is more balanced and contains samples of each of the 24 chords. The 60-chord data set is much less balanced, with some chords having no samples at all; however, it allows for more varied harmonies to occur. The data set was further segmented into groups based on tonality with subsets corresponding to major and minor keys, as well as a mixed data sets.

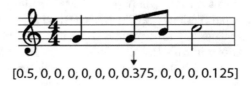

$$[0.5, 0, 0, 0, 0, 0, 0, 0.375, 0, 0, 0, 0.125]$$

Fig. 4. Example of measure and the resulting note vector

The melody from each measure was transformed into a 12-dimension *note vector* representing the relative weight of each chromatic pitch class in that measure. The weights were calculated using duration of each pitch class normalized using the reciprocal of the song's time signature. An example of this transformation is shown in Fig. 4.

The music21 toolkit [5] was used to process the data set and apply the note vector transformation. Each lead sheet was flattened into consecutive measures containing melody and chord information and information about key and time signature was extracted. In order to create example and target pairs for machine learning each measure in each of the songs should contain only one chord. For the songs where the occasional measure has more than one chord, all but the final chord in the measure are ignored. For measures with no chord notation, the chord from the preceding measure is repeated, and measures with no notes were ignored.

Training data for the ML models consisted of overlapping 8-measure sequences from the data set extracted with a step size of 1 measure. In the mixed data set there are 47564 chord and note vector pairs including "end token measures" that represent the end of the current song and the beginning of the next. Their corresponding note vectors contain only zeros. An 8 measure long window slides over the data at 1 measure steps, creating a total of 47556 sequences containing 8 measures each.

3.2   Hidden Markov Models

HMMs have been applied to a range of sequence learning problems including speech processing [1,23]. HMMs model the relationship between a "hidden" underlying process that governs an observable sequence. The hidden process is assumed to conform to the Markov property, namely that future states are dependent only on the current state, and the observed emissions of each state occur independently of their neighbouring states. For the harmony prediction problem, the observable sequence is the notes found in each measure, represented by the note vectors described above. The process which governs these observable sequences, and which we want to model, is the chord progression. Training the HMM for the harmony prediction task consists of calculating the transition (A), start (π) and emission (B) probabilities for each state using the information found in the lead sheets. The start, transition and emission probabilities can be calculated directly by counting occurrences of chord transitions and the observed notes while traversing each measure of each lead sheet.

3.3   BLSTM

The bidirectional RNN [21], as the name implies, combines an RNN which loops through the input sequence forwards through time with an RNN which moves through the input from the end. Thereby, bidirectional networks can learn contexts over time in both directions. Separate nodes handle information forwards and backwards through the network. Thus, the output at time t can utilise a summary of what has been learned from the beginning of the sequence, forwards till time step t as well as what is learned from the end of the sequence, backwards till time step t. Bidirectional LSTM was presented by Graves and Schmidhuber in 2005 [9]. This architecture could be a better choice than a typical RNN for the PSCA, chord choice can be informed both by previous chords, as well as where the harmonic progressing is going next.

The hyperparameters and structure of the BLSTM are chosen to match those used by Lim et al. in their experiments [13]. As their data set and preprocessing choices are similar to ours it is assumed that these hyperparameters will yield similar results for our implementation. The input and output layers are time distributed layers, they apply the same fully-connected layer of weights to each time step. The input layer has 12 units which represent the 12 notes, and the output layer has one unit for each of the unique chords. The hidden layers are two bidirectional RNNs with 128 LSTM units each, with hyperbolic tangent activation and a dropout rate of 0.2 between them. Softmax activation is used on the output layer in order to generate the probabilities of each chord. Categorical cross entropy is used as the cost function and the Adam optimiser is used for training with parameters following Kingma et al. [11]. The BLSTM model is trained with a batch size of 512 and early stopping to end training when validation accuracy does not improve for 10 epochs. The trained weights of the model that achieves the best validation accuracy is saved and can be loaded into the PSCA system to generate chord predictions on-the-fly. When the BLSTM model has been fitted to the data we can sample from the probability distribution of the softmax output layer to generate predictions.

3.4   Generating Harmony Prediction

In order to generate predictions from the trained HMM the Viterbi algorithm [8] is used to decide the most likely hidden sequence for a given sequence of observations. Given a model, $HMM = (A, B, \pi)$, consisting of the bigram for transitioning between chords (A), the emission probabilities (B), and the start probabilities (π), and a set of observations, O, the Viterbi algorithm begins by looking at the observations from left to right. For each observation, the max probability for all states is calculated. By keeping a pointer to the previously chosen state the algorithm can back trace through these pointers to construct the most likely path. Note that this process is deterministic, that is, a given observation sequence will always yield the same predicted hidden sequence.

Generating predictions from the BLSTM requires us to sample stochastically from the softmax output of the model at the final time step. If instead one were to choose the token with the highest probability each time, (greedy sampling) the generated chord progression can become uninteresting and repetitive. It is useful to be able to control the amount of randomness when sampling stochastically. This randomness factor is referred to as the sampling *temperature*. When choosing the next token from the softmax probability distribution the values are first reweighted using the temperature, creating a new probability distribution with more or less randomness. The higher the temperature, the higher the entropy of the resulting probability distribution.

Exactly what temperature to use when reweighting the softmax output depends heavily on the task at hand. During model evaluation a greedy sampling strategy was used, but when implementing a model for use in the PSCA system we experimented with different temperatures in order to generate less repetitive chord progressions (See Table 1).

Table 1. Example of predictions sampled at different temperatures

Original chord sequence	C	Em	Am	Am	C	Em	Am	Am
Sampled at high temp 1.1	Dm	B	G	Am	G	G	C	C
Sampled at low temp 0.1	C	G	Am	Am	G	Em	C	C

4   Evaluating the Models

To evaluate the HMM and BLSTM models, we performed a quantitative analysis of the models using cross-validation, and a qualitative analysis on the confusion matrices of these generated models to explore the accuracy of their predictions. These results were compared with the accuracy of a naive predictor that simply outputs the major triad of the first detected note within each bar. Only 12 possible chords can be generated using this approach, regardless of the song key. Namely, the major triads: *C, C#, D, D#, E, F, F#, G, G#, A, A#, B*. The accuracy of this naive prediction approach is calculated by looking at the number of correct matches between the true chords and the chords chosen using only the first note in each measure, then normalised using the total number of measures in the data. The accuracy of the naive approach on the mixed data set is 26.97%.

4.1 Cross Validation

The HMM showed signs during training of being quite sensitive to the data it was trained on, resulting at times in extremely repetitive predictions. In order to examine this more closely we apply k-fold cross validation for both models and all data sets. When applying k-fold cross validation, the data is split into k sets, one set is used for testing the model's accuracy and the remaining $k - 1$ sets are used for training the model. This process is repeated for each of the k sets and the accuracy scores are averaged. The test was run on all versions of the data set, both the 24-chord and 60-chord version, as well as on the minor key, major key and mixed key versions (as described in Sect. 3.1).

Table 2. Average accuracy for k-fold cross validation with $k = 10$ for all six datasets, a naive predictor is included for comparison. The BLSTM performed best overall.

Data set	HMM (%)	BLSTM (%)	Naive (%)
Major 24-chord	27.10	47.56	
Minor 24-chord	23.43	36.30	
Mixed 24-chord	24.38	44.53	26.97
Major 60-chord	30.56	47.27	
Minor 60-chord	17.62	36.24	
Mixed 60-chord	23.70	44.60	

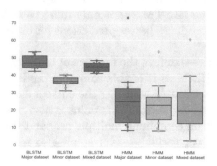

(a) K-fold cross validation accuracy 24-chord data set

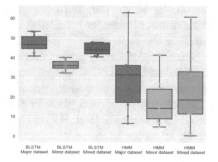

(b) K-fold cross validation accuracy 60-chord data set

Fig. 5. K-fold cross validation accuracy using 24 and 60-chord data sets. The BLSTM model had the highest median accuracy over all tests, it also had a much narrower interquartile range than the HMM models, suggesting that it is less sensitive to changes in the data set.

The results of the k-fold cross validation tests, with $k = 10$, are presented in Table 2. The BLSTM achieved higher average accuracy on all data sets. Using

24-chord classification in favour of 60-chord classification seems to result in some improvement for the HMM accuracy while the accuracy of the BLSTM predictions are strikingly similar regardless of the task requiring classification of 24 chord types, or using the imbalanced 60 chord type data set.

The accuracy for each test has been plotted using combined swarm- and box plots in Fig. 5. This shows that the accuracy of the HMM varies much more widely between folds than the accuracy of the BLSTM, regardless of using the 60-chord data sets or the 24-chord data set. This causes the interquartile range of the HMM results to be much wider, and the minimum and maximum accuracy vary between very low values—less than 10% accuracy on some folds—and higher values, over 60% accuracy, on others. This confirms the initial impressions that the HMM accuracy is strongly dependent on what segment of the data set was used to generate the transition, start and emission matrices. From the above results it is clear that the BLSTM is more robust across minor, major and mixed modes as well as having higher overall accuracy. The variations in HMM accuracy during k-fold cross validation shows how different segments of the data sets greatly affect its accuracy. In contrast, the BLSTM is more invariant to the different folds of the training data.

4.2   Creative Chord Predictions

Cross validation asks how accurate the chord choice was compared to the data set, but it is problematic that the naive predictor produces fairly good scores compared to the HMM model in particular. Figure 6 shows the confusion matrices for models trained on the 24-chord data set. We can see that the HMM models are heavily biased to return the basic chords I, IV, and V for major keys, and Im, V, VI for minor. In comparison, the BLSTM is not only more robust across minor and major tonalities, but also produces better predictions for other chords in the data set.

The model accuracy and confusion matrices confirm the findings of Lim et al. [13], in that the BLSTM outperforms the HMM as it is able to model the long term dependencies in western music more accurately, allowing the current state to be affected by several preceding and following states. This point is also presented by Raczynski et al. [19] in their work on combining multiple probabilistic models in order to encompass several musical variables. Lim et al. also mention an imbalance in their data set: "*Moreover, the fact that the training data contains more frequent occurrences of C, F and G chords (over 60% in total samples) reduced the accuracy of the HMM model which uses the prior probability to obtain the posterior*" [13, p. 4]. The imbalance between classes C, G, F and the rest of the chords is mirrored in our data set as well. Our findings show that this imbalance also affects the BLSTM model, although not as severely. The predictions generated by the BLSTM model during k-fold validation shows how the BLSTM's predictions mirror the true chord distribution of the 24-chord data set. If almost 60% of our samples belong to classes C and G, the models can simply predict only these classes for all samples and achieve somewhat reasonable accuracy, a problem also known as the accuracy paradox.

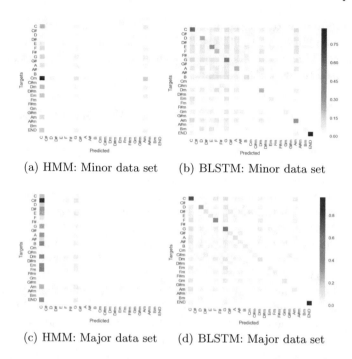

(a) HMM: Minor data set   (b) BLSTM: Minor data set

(c) HMM: Major data set   (d) BLSTM: Major data set

Fig. 6. Confusion Matrix for HMM and BLSTM using data set with 24 unique chords. These results show that the HMM misclassifies most samples as belonging to classes C, G or Am while the BLSTM has better classifications for many other chords, shown by the clear diagonal line.

There are several ways to combat the issue of imbalanced training data: Firstly of course, attempt to collect more data. Unfortunately, this is not a simple task due to the lack of a large, shared research database for popular music and jazz in lead sheet formats. An alternative could also be to generate more samples by writing music that contains the often unused chords, though naturally this would be a time-consuming task. A possible strategy would be to use Cohen's Kappa [4] in order to normalise classification accuracy using the imbalance of classes in the training samples. This approach may result in accuracy measurements that reflect the misclassifications of the minority classes better. An alternative option would be to use penalised models to impose additional cost on misclassification of a minority class. This can be used to 'force' the model to pay closer attention to the classes with the fewest training examples. Similarly, AdaCost [7], a variation of the AdaBoost method which uses the cost of misclassifications to update the training distribution on successive boosting rounds may also aid in reducing misclassification of minority classes.

Resampling the data is also a common strategy for handling unbalanced data sets, and this is the strategy used for the PSCA as a starting point for improving our results. Resampling involves changing the data set either by adding samples to (oversampling) or removing (undersampling) from the training data. In this work, undersampling was used by splitting the data into minor and major sets. Also, undersampling was used to create a small, hand-picked data set for training the HMM used in the PSCA system.

5   What Is an Accurate Chord Prediction Anyway?

It should be noted that the accuracy achieved by these models would not necessarily be considered good in other sequence prediction tasks. As mentioned in Sect. 1, there are few wrong answers when predicting chords for a given melody; several chords may sound equally good, and different listeners may appreciate different choices. Additionally, we consider predicted chords from both HMM and BLSTM models for a simple melody in Fig. 7. While the BLSTM predicts chords that go well with the melody, its accuracy for this example is only 25%. This calls into question whether predictive accuracy is an appropriate measure of quality for harmony prediction.

ML models for the creative arts must take into account that the quality of a choice could depend on the observer, as pointed out by Schmidhuber [20]. Using classification accuracy as a metric for evaluating the models, or indeed training them, may therefore be misleading; rather, we should look at different approaches to measuring musicality. One option might be to include novelty, or curiosity [20], as a training heuristic, similar to those used in reinforcement learning [3,12]. For the BLSTM, this could be incorporated into the cost function used at each training step and help reward the model for making interesting and suitable, but not exactly correct, chord choices.

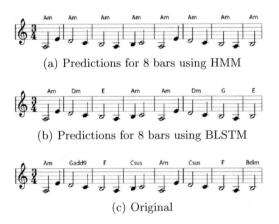

(a) Predictions for 8 bars using HMM

(b) Predictions for 8 bars using BLSTM

(c) Original

Fig. 7. Examples of different chord sequences as predicted by the models during k-fold cross validation using the 60-chord minor key data set, as well as the original chord sequence. While the BLSTM predicts suitable chords for each bar the HMM selects the root and does not generate a chord progression.

In fact, we took a heuristic version of this approach at the model selection level, by choosing models for the PSCA looper that produced the most interesting chord choices. Rather than selecting the model with the highest accuracy we chose the model that had the most aesthetically pleasing results (to our ears). For the HMM an under-sampling technique was used, creating a small hand-picked subset of the lead sheets. The BLSTM was trained using the 60-chord version of the minor-only data set and was encouraged to over fit to the training data slightly. We also used a higher sampling temperature during prediction to emphasize less likely chord choices. Examples of chord sequences produced by these two systems are shown in Fig. 8.

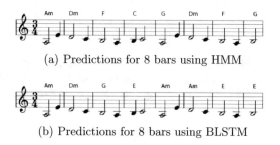

(a) Predictions for 8 bars using HMM

(b) Predictions for 8 bars using BLSTM

Fig. 8. Examples of chord sequences as predicted by the hand-tuned models trained for use in the PSCA looper. Here, the HMM's result (a) has improved significantly.

These choices seem to be more artistically rewarding. In previous work [24], two musicians explored the PSCA looper comparing the heuristically selected BLSTM and HMM models[2]. Though the heuristic choices in training served to somewhat close the gap between the HMM and BLSTM performance (as exemplified in Fig. 8), the BLSTM predictions were preferred by the participants in most sessions. While rewarding curiosity has only been implemented heuristically so far, in future work we feel that BLSTM architecture with curiosity integrated into the cost function could lead to models which better learn to generate creative and interesting predictions for artistic tasks such as harmony prediction.

6   Conclusion

In this paper we have compared two ML models of harmony prediction and discussed how they can be applied in an interactive audio looper for songwriting. The BLSTM produced better accuracy results during k-fold cross validation and exhibited more robustness against variations in the data. Although the predictions generated by the BLSTM were noticeably better than those generated by the HMM there was still an unwanted bias towards the majority classes C and G. We found that using the models that achieved the best accuracy in the PSCA

---

[2] Performance with the PSCA Looper (Direct download): https://goo.gl/59kVko.

would thereby lead to repetitive predictions with a strong major feel. This outcome was not desirable in the PSCA system. If the predicted chords are too repetitive this would be uninteresting for the performer. Similarly, predicting only major chords cause the harmonies to be quite limited. In order to mediate this issue a heuristic approach to model selection was taken, inspired by the concept of rewarding curiosity in chord prediction.

The PSCA system presents an example of an interactive system which uses machine learning techniques to enhance the musical experience of the user and has shown potential for commercial use and in live performance. In this work we have shown that finding a balance between predictions that are accurate (in reference to the training data) yet still interesting and creative is a core challenge of implementing ML in interactive music systems. Discovering efficient strategies for evaluating the models beyond their accuracy on the training data will therefor be an important direction for future research into interactive musical AI.

Acknowledgment. This work was supported by The Research Council of Norway as a part of the Engineering Predictability with Embodied Cognition (EPEC) project, under grant agreement 240862 and through its Centres of Excellence scheme, project number 262762.

References

1. Bahl, L.R., Jelinek, F., Mercer, R.L.: A maximum likelihood approach to continuous speech recognition. In: Readings in Speech Recognition, pp. 308–319. Elsevier (1990)
2. Brunner, G., Wang, Y., Wattenhofer, R., Wiesendanger, J.: Jambot: music theory aware chord based generation of polyphonic music with LSTMs. In: 2017 IEEE 29th International Conference on Tools with Artificial Intelligence (ICTAI), pp. 519–526. IEEE (2017). https://doi.org/10.1109/ICTAI.2017.00085
3. Burda, Y., Edwards, H., Pathak, D., Storkey, A., Darrell, T., Efros, A.A.: Large-scale study of curiosity-driven learning. In: Proceedings of the International Conference on Learning Representations (ICLR) (2019). https://arxiv.org/abs/1808.04355
4. Cohen, J.: A coefficient of agreement for nominal scales. Educ. Psychol. Measur. **20**(1), 37–46 (1960). https://doi.org/10.1177/001316446002000104
5. Cuthbert, M.S., Ariza, C.: music21: a toolkit for computer-aided musicology and symbolic music data. In: Downie, J.S., Veltkamp, R.C. (eds.) Proceedings of the 11th International Society for Music Information Retrieval Conference (ISMIR 2010), pp. 637–642. International Society for Music Information Retrieval, Utrecht (2010)
6. Eck, D., Schmidhuber, J.: Finding temporal structure in music: blues improvisation with LSTM recurrent networks. In: Proceedings of the 12th IEEE Workshop on Neural Networks for Signal Processing, pp. 747–756. IEEE (2002). https://doi.org/10.1109/NNSP.2002.1030094
7. Fan, W., Stolfo, S.J., Zhang, J., Chan, P.K.: Adacost: misclassification cost-sensitive boosting. In: Proceedings of the Sixteenth International Conference on Machine Learning, ICML 1999, vol. 99, pp. 97–105 (1999)
8. Forney, G.D.: The viterbi algorithm. Proc. IEEE **61**(3), 268–278 (1973)

9. Graves, A., Schmidhuber, J.: Framewise phoneme classification with bidirectional LSTM and other neural network architectures. Neural Netw. **5**(18), 602–610 (2005)
10. Hochreiter, S., Schmidhuber, J.: Long short-term memory. Neural Comput. **9**(8), 1735–1780 (1997)
11. Kingma, D.P., Ba, J.: Adam: a method for stochastic optimization. arXiv preprint arXiv:1412.6980 (2014)
12. Lehman, J., Stanley, K.O.: Abandoning objectives: evolution through the search for novelty alone. Evol. Comput. **19**(2), 189–223 (2011)
13. Lim, H., Rhyu, S., Lee, K.: Chord generation from symbolic melody using BLSTM networks. In: 18th International Society for Music Information Retrieval Conference (2017)
14. Martin, C.P., Ellefsen, K.O., Torresen, J.: Deep predictive models in interactive music. arXiv e-prints, January 2018. https://arxiv.org/abs/1801.10492
15. Martin, C.P., Torresen, J.: RoboJam: a musical mixture density network for collaborative touchscreen interaction. In: Liapis, A., Romero Cardalda, J.J., Ekárt, A. (eds.) EvoMUSART 2018. LNCS, vol. 10783, pp. 161–176. Springer, Cham (2018). https://doi.org/10.1007/978-3-319-77583-8_11
16. Mikolov, T., Sutskever, I., Chen, K., Corrado, G.S., Dean, J.: Distributed representations of words and phrases and their compositionality. In: Advances in Neural Information Processing Systems, pp. 3111–3119 (2013)
17. Pachet, F., Roy, P., Moreira, J., d'Inverno, M.: Reflexive loopers for solo musical improvisation. In: Proceedings of the SIGCHI Conference on Human Factors in Computing Systems, CHI 2013, pp. 2205–2208. ACM, New York (2013). https://doi.org/10.1145/2470654.2481303
18. Rabiner, L., Juang, B.: An introduction to hidden Markov models. IEEE ASSP Mag. **3**(1), 4–16 (1986). https://doi.org/10.1109/MASSP.1986.1165342
19. Raczyński, S.A., Fukayama, S., Vincent, E.: Melody harmonization with interpolated probabilistic models. J. New Music Res. **42**(3), 223–235 (2013)
20. Schmidhuber, J.: Developmental robotics, optimal artificial curiosity, creativity, music, and the fine arts. Connection Sci. **18**(2), 173–187 (2006)
21. Schuster, M., Paliwal, K.K.: Bidirectional recurrent neural networks. IEEE Trans. Sig. Process. **45**(11), 2673–2681 (1997)
22. Simon, I., Morris, D., Basu, S.: Mysong: automatic accompaniment generation for vocal melodies. In: Proceedings of the SIGCHI Conference on Human Factors in Computing Systems, CHI 2008, pp. 725–734. ACM, New York (2008). https://doi.org/10.1145/1357054.1357169
23. Tokuda, K., Zen, H., Black, A.W.: An HMM-based speech synthesis system applied to English. In: IEEE Speech Synthesis Workshop, pp. 227–230 (2002)
24. Wallace, B.: Predictive songwriting with concatenative accompaniment. Master's thesis, Department of Informatics, University of Oslo (2018)

Stochastic Synthesizer Patch Exploration in Edisyn

Sean Luke$^{(\boxtimes)}$

George Mason University, Washington, DC, USA
sean@cs.gmu.edu

Abstract. Edisyn is a music synthesizer program (or "patch") editor library which enables musicians to easily edit and manipulate a variety of difficult-to-program synthesizers. Edisyn sports a first-in-class set of tools designed to help explore the parameterized space of synthesizer patches without needing to directly edit the parameters. This paper discusses the most sophisticated of these tools, Edisyn's Hill-Climber and Constrictor methods, which are based on interactive evolutionary computation techniques. The paper discusses the special difficulties encountered in programming synthesizers, the motivation behind these techniques, and their design. It then evaluates them in an experiment with novice synthesizer users, and concludes with additional observations regarding utility and efficacy.

Keywords: Synthesizer patch design ·
Interactive evolutionary computation

Edisyn is Java-based open-source patch editor library for hardware and software music synthesizers. A *patch* is the traditional term for a particular set of parameters which, when programmed into a music synthesizer, enables a performer to play a particular sound. That is, a patch is the music synthesizer equivalent of setting certain stop knobs or drawbars on an organ. Many synthesizers have poor interfaces or are otherwise difficult (and in some cases impossible) to program directly, making attractive the notion of a *patch editor*, piece of software which enables the musician to more easily program these devices remotely via his laptop. Edisyn provides patch editors for a variety of difficult-to-program synthesizers. I am Edisyn's software developer.

Edisyn's particular strength lies in assisting the musician in exploring a synthesizer's parameter space. Indeed, I believe Edisyn is easily the most capable general-purpose synthesizer patch editor available in this respect. This paper discusses two of the most sophisticated patch-exploration tools in Edisyn's toolbox: its stochastic hill-climber and constrictor facilities. Both of these tools apply an interactive version of evolution strategies (ES) to do this assisted exploration; but the two tools have very different philosophies and heuristic approaches.

The hill-climber and the constrictor are both evolutionary computation (EC) methods, treating candidate synthesizer patches as the individuals in question. EC has been used for at least 25 years [1] to search the space of synthesizer

© Springer Nature Switzerland AG 2019
A. Ekárt et al. (Eds.): EvoMUSART 2019, LNCS 11453, pp. 188–200, 2019.
https://doi.org/10.1007/978-3-030-16667-0_13

patches, but the lion's share of this literature has been in *evolutionary resynthesis*, where an automated function assesses candidate patches based on desired predefined qualities. In contrast, Edisyn's hill-climber and constrictor are examples of *interactive EC* [2], where the assessment function is a human.

Interactive EC is not easy: humans are fickle, arbitrary, and easily bored. Whereas a typical (non-interactive) EC run may consider many thousands of candidate solutions, a human cannot be asked to assess more than a few hundred candidates before he gives up. Unfortunately, because synthesizer patches are high dimensional (10–1000 parameters) and often complex, making progress in this space with a small number of presentations is hard. This problem is commonly known as the *fitness bottleneck* [3] in the interactive EC literature. Humans also have noisy preferences and often change their minds, and so while one might imagine that interactive EC would enable a human to search for a target sound fixed in his mind, in fact such tools are perhaps best used to help him explore the space. When the user comes across an "interesting" new sound, he may direct the system to start searching in that *direction* rather than towards some predefined *goal.*

The space of synthesizer patches presents its own special domain-specific difficulties. Patches take up time to be auditioned and heard, and hardware synthesizers are slow in receiving them over MIDI (the standard protocol for patch transfer): this restricts the rate at which patches may be presented to the user. Additionally, while synthesizer patches are typically fixed-length arrays (and Edisyn assumes this representation), different synthesizers have parameter spaces with a wide range of statistical linkage, numerosity, resolution, and degree of impact on the final sound. At present Edisyn supports almost 30 patch editors for 15 different synthesizers covering many synthesis modalities, including additive, subtractive, and FM synthesis, plus samplers and ROMplers. Creating a single general-purpose tool to explore all these spaces is not easy.

Finally, patch parameter types are heterogeneous, taking three forms. First there are *metric* parameters, normally integers, such as volume or frequency. Second, there are *non-metric* (or categorical) parameters, such as choice of oscillator wave type. Third, *hybrid* parameters have *both* metric and non-metric ranges: for example, a MIDI channel parameter may be set to 1...16, or to Off, Omni, MPE High/Low, etc. A general-purpose patch exploration facility ought to be able to accommodate any combination of these. (Incredibly, there also exist synthesizers with holes in their valid parameter ranges.)

1   Previous Work

The interactive synthesizer patch exploration literature has considered different ways to tackle some of these challenges. [4] focuses on interfaces designed to speed the assessment and selection of solutions, albeit with very small parameter spaces. To simplify the search space, [5] updates a parameterized model instead of a sample (essentially a kind of estimation of distribution algorithm). [6] essentially rounds candidate solutions to the nearest known patch drawn from a library.

Fig. 1. Edisyn's Yamaha DX7 patch editor window, showing the tab for FM Operators 1 and 2.

Best known is MutaSynth, a metaheuristic patch-exploration tool which found its way into a commercial product (Nord's Modular G2 synthesizer editor) [7–9]. MutaSynth is manual in operation: the user selects a patch or two from a palette of "best so far" sounds, then chooses how to recombine or mutate them. The system then produces several recombined or mutated children, and the user decides whether to add any children to the palette. This requires the user to make several decisions per iteration, but ideally the "best so far" palette would maintain user interest and increase diversity, thus helping avoid getting stuck in local optima.

Curiously, nearly all previous work (including MutaSynth and similar methods [10,11]) problematically assumes that *all* parameters are metric, either by restricting them to be so or by shoehorning non-metric parameters into metric spaces. But often as many as half of the parameters in a typical synthesizer patch are non-metric or are hybrid.

While Edisyn and most other examples from the literature focus on fixed-length arrays (as the large majority of synthesizers use them), some recent interactive EC literature has begun to examine unusual variable-sized representations for modular synthesizers [12] or custom neural-network based synthesis techniques [13,14].

2   Edisyn

Edisyn supports a variety of synthesizers, but places a special emphasis on synthesizers with difficult interfaces or difficult-to-program architectures. Of particular interest to discussion here are FM and additive synthesizers. Additive synthesizers are notorious for high dimensional patches: for example the Kawai K5000 has on the order of a thousand parameters. On the other hand, FM synthesizers tend to have parameters that are highly epistatic, difficult to control, and *very* difficult to predict. This is not the fault of the devices, but rather is due to the nonlinear nature of the FM synthesis process. Figure 1 shows one pane of Edisyn's editor for the Yamaha DX7, a famous FM synthesizer.

Like many patch editors, Edisyn provides many features to assist in directly programming these synthesizers. But Edisyn also provides a large assortment of tools to partially automate the search for patches of interest. Simple Edisyn tools include *weighted patch randomization*, where a patch is mutated to some degree; and *weighted patch recombination*, where two patches are merged, with an emphasis placed on one patch versus the other. Additionally, Edisyn's *nudge* facility allows the user to use recombination to push a patch towards or away from up to four different target patches, not unlike mixing paints on a palette.

Algorithm 1. Mutate($P, v, 0 \leq weight \leq 1$)

Input: Parameter P with current value v, *weight*
if P is metric **then**
    return MetricMutate(P, v, *weight*)
else if P is a hybrid parameter and v is a metric value **then**
    if coin flip of probability 0.5 **then**
        return MetricMutate(P, v, *weight*)
    else if coin flip of probability *weight* **then**
        return a random non-metric value in P
else if P is a hybrid parameter and v is not a metric value **then**
    if coin flip of probability *weight* **then**
        if coin flip of probability 0.5 **then**
            return a random metric value in P
        else return a random non-metric value in P
else if coin flip of probability *weight* **then**
    return a random metric value in P
return v

Algorithm 2. MetricMutate($P, v, 0 \leq weight \leq 1$)

Input: Parameter P with current value v; *weight*
$q \leftarrow weight \times (1 + \text{metric max of } P - \text{metric min of } P)$
$l \leftarrow \max(v - \lfloor q \rfloor, \text{metric min of } P)$
$h \leftarrow \min(v + \lceil q \rceil, \text{metric max of } P)$
return a uniform random selection from $[l, h]$

The above tools rely in turn on three mutation and recombination algorithms to do their work. Each of these algorithms takes a single weight as a tuning parameter, to minimize the complexity of the system for the user. They also take into consideration the fact that on effectively all synthesizers, metric parameters are integers (due to MIDI). The algorithms are summarized below, with details in Algorithms 1–4:

Mutate (Algorithm 1) mutates a parameter. The weight defines the probability that the value is randomized (if non-metric) and/or the degree to which it will be modified (if metric). If the parameter is hybrid, with 0.5 probability its value will be mutated from metric to non-metric (or vice versa). Weighted patch randomization and nudging both use this operator.

Recombine (Algorithm 3) modifies a parameter in one patch to reflect the influence of another patch. The weight defines the probability of recombination occurring, either via weighted interpolation (if metric) or choosing one value or the other (if non-metric, or if the parameter is hybrid and one value is metric while the other is not). Weighted patch recombination and nudging (towards a target) both use this operator.

Algorithm 3. Recombine($P, v, w, 0 \leq weight \leq 1$)

Input: Parameter P with current values v, w; *weight*
if coin flip of probability *weight* **then**
    if v and w are both metric values in P **then**
        return a uniform random selection from $[v, w]$
    else if coin flip of probability 0.5 **then**
        return w
return v

Algorithm 4. Opposite($P, v, w, 0 \leq weight \leq 1$, boolean *flee*)

Input: Parameter P with current values v, w; *weight*; *flee*
if $v = w$ and *flee*=true **then**
    if v is non-metric **then**
        if coin flip of probability *weight* **then**
            return a random non-metric value from $P - \{w\}$
    else
        return $v + 1$ or $v - 1$ at random as constrained by the metric min and metric   max of P
else if both v and w are metric values in P **then**
    $q \leftarrow \lceil v + weight \times (v - w) \rceil$
    $q \leftarrow \min(\max(q, \text{metric min of } P), \text{metric max of } P)$
    return a uniform random selection from $[v, q]$
return v

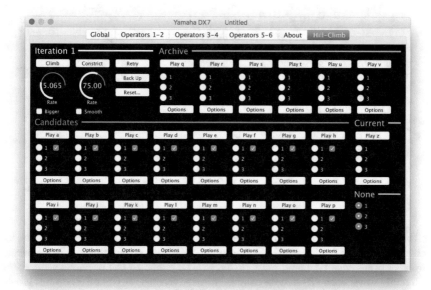

Fig. 2. Edisyn's Hill-Climber.

Opposite (Algorithm 4), an unusual recombination variant, finds a value on the *other side* of parameter P from Q, notionally because we had moved from Q to P and wished to continue in that direction. The weight determines the degree to which the value moves away from Q if both are metric. If either value is non-metric, or if they are equal, then P's value stays as it is.

A subtly different form of Opposite is used when we want P to *flee from* Q: here, if the two values are equal, then if they are non-metric P's value is randomly made different from Q with a probability determined by the weight; else if they are metric, P's value is nudged slightly away from Q (so that $|P - Q| > 0$ and so future Opposite-recombinations can push it further away). Nudging (away from a target) uses the flee version of this operator.

3   Stochastic Patch Exploration Tools

Edisyn also has two stochastic patch exploration procedures: the Hill-Climber and the Constrictor. Both procedures are available in Edisyn's Hill-Climber panel, shown in Fig. 2. The purpose of these methods is to help the user to rapidly explore the patch parameter space in search of patches interesting or useful to him; after discovering such patches in rough form, the user can then fine-tune the patches by hand in Edisyn's editor.

The two procedures are both variations of interactive evolution strategies, but have different goals and heuristics, and consequently very different dynamics. The Hill-Climber applies an ES with a specialized breeding procedure to

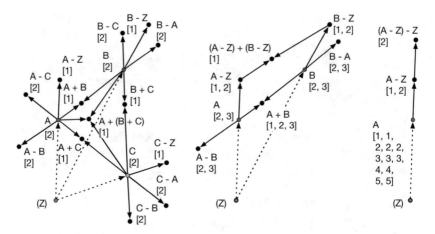

Fig. 3. Hill-Climber breeding procedures when the user selects three parents (left), two (center), or just one (right). In these diagrams, A, B, and C denote those three parents in order of preference (A being most preferred). (Z) denotes the previous-generation's A. All nodes other than (Z) denote children created from A, B, C, and Z: each is labeled with the operation used to generate the child, such as $B - A$. The operation $X + Y$ is recombination (preferring X) and $X - Y$ is non-flee opposite-recombination. Each child is then mutated at least once, or multiple times as indicated in square brackets: e.g. [3] is three mutations in series. When multiple numbers are in brackets, this indicates that multiple children are created using this procedure, with different numbers of mutations. For example $A - Z$ [1, 2] means that two children are created using $A - Z$, with one child mutated once, while the other is mutated twice.

wander through the space guided by the user's preferences as a fitness function. In contrast, the Constrictor applies an ES with crossover only, again guided by the user's preferences, to gradually reduce the convex hull or bounding box of the search space until it has (intentionally prematurely) converged to a point in the space. The two procedures again rely on the above Mutate, Recombine, and Opposite operators (in non-flee form). The Constrictor may also use an additional *Crossover* operator, described later.

Like other Edisyn patch-exploration tools, the Hill-Climber and Constrictor are general-purpose. The two methods are designed to tackle the fitness bottleneck problem in two ways. First, the techniques both act essentially as black boxes, requiring only a single user-selectable weight parameter each, so as not to overwhelm the user with complexity, and to expedite selections and decisions. Second, both try to avoid poor-quality regions of the space so as to speed search for viable candidates: the Hill-Climber does this by making largely local moves from earlier-chosen individuals; and the Constrictor tries to reduce the space to just the convex hull or bounding box of well-vetted solutions.

4   The Hill-Climber

The *Hill-Climber* is essentially a (μ, λ) evolution strategy variant which emphasizes local search. Rather than have the user specify how to generate children, the Hill-Climber applies a custom breeding protocol, shown in Fig. 3, designed to sample from a distribution whose size is proportional to the diversity in the user's choices (a notion drawn from Differential Evolution). The protocol also applies the heuristic that if the user has been choosing candidates in a certain direction, he may wish to consider some candidates *even further* in that direction. Finally, the protocol tries to balance both diversity-creating operations and recombination operations.

Algorithm 5. Crossover($P, v, w, 0 \leq weight \leq 1$)

Input: Parameter P with current values v, w; *weight*
if coin flip of probability *weight* **then**
   if coin flip of probability 0.5 **then**
      return w
return v

Though the user is not required to do so by default, Edisyn permits the user to bias this search by nudging candidates toward desired targets, editing them, backing up, storing and recalling search points, and constraining the parameters being explored (a simple form of dimensionality reduction). The user has control over a single parameter, the *mutation weight*, which affects both the probability of mutative changes and their strength. Increasing this weight will increase variance among patches, while decreasing it leads to convergence. The recombination weight is fixed, as I have found anecdotally that control over this parameter is confusing to patch designers and is rarely changed.

Edisyn's Hill-Climber initially builds 16 children mutated from the original patch, which are then sent to the synthesizer to be auditioned in turn. The user iteratively selects up to three preferred "parent" patches chosen from the 16, plus optionally the original patch and an archive of up to six patches he has collected over time. After this, he Hill-Climber generates 16 new children from the parents, auditions them, and so on. In EC this would be, loosely, an approximately *(3, 16)* evolution strategy. While 16 is the default, the Hill-Climber can also be set to use a population size of 32.

To generate children, the Hill-Climber employs three different but fixed breeding strategies depending on the number of parents chosen, as shown in Fig. 3. The strategies rely only on the (up to) three selected parents in order, plus the previous #1 parent. The operations are meant to provide both mixtures of the parents as well as diversity from them, and to heuristically continue moving in the direction the user had been previously going: a form of momentum. As detailed in Fig. 3, all the breeding procedures ultimately rely on the aforementioned mutation, recombination, and (non-flee) opposite-recombination procedures.

5   The Constrictor

Edisyn's *Constrictor* also applies an evolution strategy but with a different goal. Whereas the Hill-Climber behaves as a kind of semi-directed random walk through the space, the Constrictor begins with several different pre-defined patches and slowly reduces their convex hull or bounding box through repeated selection and recombination. The idea behind the Constrictor is to reduce the space to patches whose parameters are combinations of well-vetted patches and so are (hopefully) more likely to be useful in the first place.

The Constrictor works as follows. First, some $N \leq 16$ individuals are loaded from user-selected patches, and the remaining $16 - N$ individuals are generated from the original N by iteratively selecting two of the N and recombining them to create a child. These 16 individuals then form the initial population. The constrictor then iterates as follows. The user first selects some $M \ll 16$ individuals from this group that he *doesn't* like, perhaps because they are of poor quality, or are uninteresting, or are duplicates of others. The system then replaces these M individuals with new children recombined from the remaining $16 - M$, and the iteration repeats. There is no mutation: at the limit this approach will converge to a single patch due to, effectively, the premature convergence of crossover. Again, while a population size of 16 is the default, Edisyn can be set to use 32 instead.

Depending on user settings, the Constrictor uses one of two different recombination approaches: either it uses *Recombine* (Algorithm 3, described earlier) or it uses *Crossover* (Algorithm 5). Whereas Recombine directly crosses over non-metric parameters while doing a weighted randomized average of metric parameters, Crossover instead is a traditional crossover procedure: it directly crosses over *all* parameters. Just like Recombine (and other operators), Crossover takes a single tuning parameter. When used with Recombine, the Constrictor is known as a *Smooth Constrictor*, and when used with Crossover, it is known as an *Unsmooth Constrictor*.

6   Experiment

It is nontrivial to formally analyze or assess a system with a human in the loop whose mind changes rapidly. Furthermore, while Edisyn's Hill-Climber and Constrictor can be used to optimize for a target sound, they are designed for, and tuned for, assisted patch *exploration* due to the human factor. Thus we attempted to gauge user response to these tools, and to solicit their opinion as to the tools' effectiveness. We asked a sample of 28 university-level computer science or music technology students to assess different Edisyn patch-exploration techniques. Almost all of the subjects also had significant music performance experience, but only a handful had previous experience with music synthesizers.

Choice of Synthesizer. We chose the Yamaha DX7, the archetypal FM synthesizer, as our target platform. FM synthesis presents an obvious demonstration of why patch-exploration tools are useful: it is counterintuitive and difficult, but at the same time is very widely used. Though they are general-purpose, Edisyn's techniques were developed originally for FM synthesis devices because of these reasons. This isn't unusual: indeed the first application of evolutionary computation to patch design [1] also targeted FM as well, and with a similar justification. We also felt that FM's difficulty in patch design (for everyone) would also put all but the very most experienced patch designers on roughly equal footing in the study.

Procedure. We began the study with an hour-long hands-on tutorial on synthesizers, FM synthesis, and how to program DX7 patches in Edisyn. The objective was to get subjects up to the level of a novice DX7 programmer. We also included a short introduction in how to use the Hill-Climber and Constrictor features. Each subject used his own laptop, working on Edisyn to control Dexed, a well-known open source emulator of the DX7 (a DX7 itself could easily have been used as well). The DX7's pitch-modification parameters (*Transpose* and the *Pitch Envelope*) were disabled, as were the *Fixed Frequency* and *Frequency Fine* parameters for each DX7 operator, so as to eliminate the highly sensitive, unhelpful, and often atonal swings associated with them.

We then asked each subject to try four different techniques (direct programming, hill-climbing, smooth constrictor, and unsmooth constrictor, in randomized order) to explore the patch parameter space in search of sounds which were in their opinion musical, interesting, or useful. Each trial began with a specific initial patch (or as set of four patches for the constrictors), from which the subjects would begin to explore. In the case of direct programming, the subject began with the initial patch and explored the space by directly modifying the parameters using Edisyn's primary features (including Undo). A trial lasted approximately five minutes, and after each trial the subject provided his opinions of the technique, both as ratings from 1 to 10 and in text. There were eight trials in all (two sets): the first set of trials were a warm-up period and intentionally discarded. After all trials had concluded, we discussed the subject's opinions with them.

Results. The results to the primary question ("On a scale from 1 to 10 (higher is better), rate your satisfaction with this method.") were as follows:

Method	Mean	Variance
Direct Programming	5.857143	4.9417990
Hill-Climbing	**7.107143**	3.0621693
Smooth Constrictor	**7.535714**	1.7394180
Unsmooth Constrictor	**7.035714**	2.2579365

Boldface indicates methods which were statistically significantly preferred over Direct Programming, using an ANOVA at $p = 0.05$ with a Holm-Bonferroni post-hoc test. The three other methods had no statistically significant difference among them, though Smooth Constrictor was very close to being preferred over all others.

These results were corroborated by subject discussion after the trials: with few exceptions, the subjects found the exploratory methods much more effective than direct programming. Subjects also generally noted a preference for Smooth Constrictor, particularly over its Unsmooth counterpart. Smooth was preferred over Unsmooth largely because Unsmooth created too many unusual and deviant sounds.

In an attempt to verify why certain methods were preferred over others, the study also asked several follow-up questions asking which methods were more effective at unusual sounds vs. ones more useful in a compositional setting, or methods which deviated the most rapidly from the initial settings. The results for these questions were not statistically significant.

7   Anecdotal Observations

Due to the difficulty in obtaining experts, the previous experiment was largely based on novices to synthesizer programming. But Edisyn's tools were really designed with experienced users in mind. I view myself as relatively experienced in patch development, and have personally found these tools to be very useful; and have observed the same in others more expert than myself. And in the case of FM, "experts" of any kind are rare.

In using the Hill-Climber in particular I have also found, not unexpectedly, that reduced dimensionality is key. Edisyn was developed with FM synthesis in mind; but high-dimensional additive synthesizers have proven the most challenging, despite their parameters' typically low epistasis. Also, changes in certain sound features (like modulation of pitch or volume) will mask subtle changes in others. For these reasons, it is critical to constrain the search to just those parameters of interest.

Another difficulty stems from large-range non-metric parameters. For example, many ROMplers have a "wave" parameter which specifies which of a great many stored samples should be played by an oscillator. These samples are arbitrarily ordered, and so this parameter is non-metric but with a large set of values to choose from. Some E-Mu ROMplers, such as the Morpheus and Ultra Proteus, also offer parameters for hundreds of arbitrarily ordered filter options. Parameters such as these are critical to a ROMpler's patch, but their complete lack of gradient presents a worst-case scenario for optimization tools. I suspect there may be no way around this.

Like most evolutionary methods, Edisyn's exploration tools can get caught in local optima. MutaSynth historically championed temporary "best so far" storage as a way to help the user maintain diversity and thus work his way out of local optima. Edisyn does this as well, sporting a "best so far" archive of six candidate patches: the user can store his best discovered candidates in this archive at any time, and during hill-climbing he can also include these candidates as part the selection procedure. However, an Edisyn user can also back up (undo) to previous iterations in the hill-climbing or constricting process, or can retry an iteration if he dislikes the results; and I have personally found myself more reliant on these two facilities than the archive facility when dealing with local optima.

Last, but hardly least: I have found the following workflow more effective than using any of these tools in isolation. Namely, one might start with a constrictor to work down to a general area of interest in the parameter space, then switch to the hill-climber to explore in that region, and then ultimately switch to hand-editing to tune the final result.

8 Conclusion and Future Work

This paper discussed the synthesizer patch editor toolkit Edisyn and described and examined its stochastic patch exploration tools. These tools are interactive evolutionary algorithms meant to assist the patch programmer in exploring the parameterized space of synthesizer patches in order to improve them and to discover new and useful ones.

Though their usefulness is obvious and their application straightforward, general-purpose evolutionary patch exploration tools are very rare. Many patch editors (and some synthesizers, such as the Waldorf Blofeld) provide crude patch randomization or recombination tools, but almost none provide more sophisticated techniques. Edisyn is notable both in its array of patch exploration tools and also in the fact that the tools are all designed to work with any patch editor written in the toolkit, so long as a patch is defined as a vector of values from heterogeneous metric, non-metric, or hybrid domains.

Because the examples are so rare, there are many opportunities in this area for exploring additional interactive metaheuristic approaches. One obvious extension of this work of particular interest to the author is into spaces with more complex evolutionary representations. There has been significant interest of late in hardware and software synthesizers consisting of many *modules*, of any number, and connected any number of ways. Simpler versions of these devices (so-called "semi-modular" synthesizers) have patches in the form of arbitrarily-long lists; more sophisticated "fully modular" architectures would require an arbitrary and potentially cyclic directed graph structure as well. I have myself written fully-modular software synthesizers. But as has long been recognized in metaheuristics, and particularly in the genetic programming and neuro-evolution subfields, optimizing such representations is nontrivial, ad-hoc, and not particularly well understood even now. Though modular synthesizers are now very popular, the literature in modular synthesizer patch exploration (notably [12]) is only in its infancy, and this area is well worth further consideration.

Hopefully in the future we will see more synthesizer patch editors with tools such as those demonstrated in this paper. Synthesizers are popular but challenging to program devices, and some synthesizer approaches (additive, FM) are notoriously so. Editors improve the user interface of the machines but do not necessarily help overcome these inherent difficulties: but patch exploration tools can make this possible even for inexperienced users.

Acknowledgments. My thanks to Vankhanh Dinh, Bryan Hoyle, Palle Dahlstedt, and James McDermott for their considerable assistance in the development of this paper.

References

1. Horner, A., Beauchamp, J., Haken, L.: Musical tongues XVI: genetic algorithms and their application to FM matching synthesis. Comput. Music J. **17**(4), 17–29 (1993)
2. Takagi, H.: Interactive evolutionary computation: fusion of the capabilities of EC optimization and human evaluation. Proc. IEEE **89**(9), 1275–1296 (2001)
3. Biles, J.A.: GenJam: a genetic algorithm for generating jazz solos. In: ICMC, pp. 131–137 (1994)
4. McDermott, J., O'Neill, M., Griffith, N.J.L.: Interactive EC control of synthesized timbre. Evol. Comput. **18**(2), 277–303 (2010)
5. Seago, A.: A new interaction strategy for musical timbre design. In: Holland, S., Wilkie, K., Mulholland, P., Seago, A. (eds.) Music and Human-Computer Interaction, pp. 153–169. Springer, London (2013). https://doi.org/10.1007/978-1-4471-2990-5_9
6. Suzuki, R., Yamaguchi, S., Cody, M.L., Taylor, C.E., Arita, T.: iSoundScape: adaptive walk on a fitness soundscape. In: Di Chio, C., et al. (eds.) EvoApplications 2011. LNCS, vol. 6625, pp. 404–413. Springer, Heidelberg (2011). https://doi.org/10.1007/978-3-642-20520-0_41
7. Dahlstedt, P.: A MutaSynth in parameter space: interactive composition through evolution. Organized Sound **6**(2), 121–124 (2001)
8. Dahlstedt, P.: Evolution in creative sound design. In: Miranda, E.R., Biles, J.A. (eds.) Evolutionary Computer Music, pp. 79–99. Springer, London (2007). https://doi.org/10.1007/978-1-84628-600-1_4
9. Dahlstedt, P.: Thoughts of creative evolution: a meta-generative approach to composition. Contemp. Music Rev. **28**(1), 43–55 (2009)
10. Collins, N.: Experiments with a new customisable interactive evolution framework. Organised Sound **7**(3), 267–273 (2002)
11. Mandelis, J.: Genophone: evolving sounds and integral performance parameter mappings. In: Cagnoni, S., et al. (eds.) EvoWorkshops 2003. LNCS, vol. 2611, pp. 535–546. Springer, Heidelberg (2003). https://doi.org/10.1007/3-540-36605-9_49
12. Yee-King, M.J.: The use of interactive genetic algorithms in sound design: a comparison study. Comput. Entertainment **14** (2016)
13. Ianigro, S., Bown, O.: Plecto: a low-level interactive genetic algorithm for the evolution of audio. In: Johnson, C., Ciesielski, V., Correia, J., Machado, P. (eds.) EvoMUSART 2016. LNCS, vol. 9596, pp. 63–78. Springer, Cham (2016). https://doi.org/10.1007/978-3-319-31008-4_5
14. Jónsson, B., Hoover, A.K., Risi, S.: Interactively evolving compositional sound synthesis networks. In: GECCO, pp. 321–328 (2015)

Evolutionary Multi-objective Training Set Selection of Data Instances and Augmentations for Vocal Detection

Igor Vatolkin[1](✉) and Daniel Stoller[2]

[1] TU Dortmund, Dortmund, Germany
igor.vatolkin@tu-dortmund.de
[2] Queen Mary University of London, London, UK
d.stoller@qmul.ac.uk

Abstract. The size of publicly available music data sets has grown significantly in recent years, which allows training better classification models. However, training on large data sets is time-intensive and cumbersome, and some training instances might be unrepresentative and thus hurt classification performance regardless of the used model. On the other hand, it is often beneficial to extend the original training data with augmentations, but only if they are carefully chosen. Therefore, identifying a "smart" selection of training instances should improve performance. In this paper, we introduce a novel, multi-objective framework for training set selection with the target to simultaneously minimise the number of training instances and the classification error. Experimentally, we apply our method to vocal activity detection on a multi-track database extended with various audio augmentations for accompaniment and vocals. Results show that our approach is very effective at reducing classification error on a separate validation set, and that the resulting training set selections either reduce classification error or require only a small fraction of training instances for comparable performance.

Keywords: Vocal detection ·
Evolutionary multi-objective training set selection · Data augmentation

1 Introduction

The goal of music classification is to assign music data to categories such as music genres, emotions, harmonic properties, and instruments. To build a model for classification, a *training set* containing music examples with manually annotated labels is normally required. The size of this training set is crucial – if it is too small, overfitting occurs, and generally prediction performance tends to increase with more training data. While the available music data has grown in recent years (for example from 215 music tracks for genre recognition [10] to 106,574 pieces in the Free Music Archive [8]), labels are still often not available in great

© Springer Nature Switzerland AG 2019
A. Ekárt et al. (Eds.): EvoMUSART 2019, LNCS 11453, pp. 201–216, 2019.
https://doi.org/10.1007/978-3-030-16667-0_14

quantity, since annotation requires too much human effort. This is also the case for vocal activity detection [28].

Because of this label scarcity problem, *data augmentation* is often used to increase the number of annotated music examples [13]. In data augmentation, new training instances are created from existing ones by introducing variations, but keeping the already existing annotation. For example, annotated music tracks for genre classification can be amplified to different loudness levels, which should not affect the genre labelling, and all variants included in the training set. Due to its popularity, frameworks for music data augmentation such as the MATLAB-based *Audio Degradation Toolbox* [16] and the *muda* Python package for Musical Data Augmentation [18] were developed to allow for an easy application of various perturbations like adding noise or pitch shifting. However, data augmentation can also hurt performance when the wrong assumptions about the classification problem are made and as a result, the augmentation rules are set-up incorrectly [27].

The arrival of *big data* to music data analysis also exacerbated another problem: Complex classification models that work well with very large data sets, such as deep neural networks, are often hard to interpret for musicologists and users, who may wish to understand the defining properties of the different musical categories the model is operating with. Model prototyping also becomes more difficult, since feedback regarding model performance is obtained only infrequently due to long training times. Most importantly, corrupted or otherwise unrepresentative instances in the training set common in large datasets can unknowingly impact performance negatively, which are hard to find in these large datasets. Usually, models are trained on the full dataset, assuming instances are independent and identically distributed, so that the presence of such outlier instances effectively imposes an upper bound on the attainable performance, regardless of model choice.

Instead of training from more and more data regardless of its quality and applying a fixed set of augmentations, one may thus consider to focus on the identification of *smart data*, i.e., observations which are particularly useful to extract the most relevant properties of the target class. In this paper, we therefore propose a novel multi-objective evolutionary framework, which optimises the selection of the training instances and augmentations for a dataset to find the best solutions trading off the number of required training instances and the resulting classification performance.

As an application scenario to validate our approach, we have selected vocal detection as a prominent and well-researched task in music classification, and previous methods for the task are discussed in Sect. 2.1. Related work on training set selection is addressed in Sect. 2.2. In Sect. 3, we briefly introduce the backgrounds of multi-objective optimisation and outline our approach. Section 4 deals with the setup of our experimental study. The results are discussed in Sect. 5. We conclude with the most relevant observations and discussion of future work in Sect. 6.

2   Related Work

We review related work in the field of vocal activity detection and in evolutionary optimisation for training set selection in Sects. 2.1 and 2.2.

2.1   Vocal Detection

Earlier approaches for singing voice detection mostly involve heavy feature engineering to enable classification [17,24,25], which yields moderate performance. However, attempting to further improve accuracy by refining the features suffers from diminishing returns, since it becomes harder to manually specify exactly which aspects of the audio data are relevant for classification.

More recently, approaches based on neural networks [26,27] have been proposed that promise to reduce the required feature engineering and instead learn the relevant features directly from the data. While this avoids performance bottlenecks due to suboptimal feature design and can theoretically deliver high accuracy, it typically requires larger amounts of labelled data. Since publicly available, labelled data for singing voice detection are limited [28], ways to prevent overfitting were proposed alongside these models: Schlüter [26] applied a convolutional neural network (CNN) on weakly labelled audio excerpts in an attempt to extend the amount of usable data for training, obtaining improved performance compared to previous work, but also finding that the network also classifies simple sinusoids with pitch fluctuations as singing, which indicates the concept of a singing voice was not learned correctly. Similarly, a joint vocal separation and detection was employed by [28] aiming to exploit both singing voice separation datasets for vocal activity detection. Schlüter et al. [27] explored the benefit of different data augmentation techniques that among other aspects vary pitch and tempo of the audio excerpts. The results were mixed – some augmentations were helpful, but some were also detrimental to performance, suggesting that both data augmentation and feature extraction require prior knowledge to decide what the classifier outputs should be invariant to, and performance depends on the accuracy of this knowledge. Furthermore, all mentioned approaches can suffer from outliers in the training data. We thus aim to automate the selection of representative training instances as well as helpful data augmentations to increase vocal detection performance, which enables more robust classification models.

2.2   Multi-objective Evolutionary Optimisation and Training Set Selection

Multi-objective optimisation evaluates solutions with regard to several optimisation criteria (see Sect. 3.1). In that case, many solutions become incomparable. As an example, consider a classification model which is fast but has a higher classification error, and another one which is slow, but has a lower error. The first one is better with regard to runtime, the second one with regard to classification error. However, it is still possible to create models which are both faster and have higher classification performance, but it becomes harder to identify the set of trade-off solutions. In this context, evolutionary algorithms (EA) [2] were considered [7]. EAs are often applied for such complex optimisation tasks, where a large search space makes it challenging to find a sufficiently good solution in acceptable time, and where other methods such as gradient descent can not be applied (e.g. for objective functions which are not differentiable or are multi-modal and have many local optima).

In music research, EAs were applied for instance for music composition [20] or feature selection [9]. A multi-objective EA with the target to minimise the number of selected features and the classification error was presented in [29]. EAs have proven their ability to generate new features for music classification by exploring nearly unlimited search spaces of combinations of different transforms and mathematical operations [15,19,22]. EAs have also been successfully applied for training set augmentation in bioinformatics, where the generation of new training data may be very expensive [11,32], and for training set selection (TSS) [6]. In [1], TSS was explicitly formulated as a multi-objective problem of simultaneously maximising the classification accuracy and the reduction in training set size. Although training set selection was already applied for classification of acoustical events [21], we are not aware of any studies on multi-objective TSS for classification of music data.

3   Approach

Our approach consists of leveraging an evolutionary, multi-objective optimisation method, which is described in Sect. 3.1, and applying it to training set selection, as shown in Sect. 3.2.

3.1   Multi-objective Evolutionary Optimisation

In the following, we formally introduce the problem of multi-objective optimisation with a focus on evolutionary methods. Let $\mathcal{X}$ be the decision or search space and $\mathbf{f} : \mathcal{X} \mapsto \mathbb{R}^d$ the vector-valued objective function with $d \geq 2$. $\mathcal{F} = \{f(\mathbf{x}) : \mathbf{x} \in \mathcal{X}\}$ is called the objective space. The goal of multi-objective optimisation is to simultaneously minimise all d dimensions $f_1, ..., f_d$ of $\mathbf{f}$ (we provide here a short definition without constraints, for the latter see [33]).

A solution $\mathbf{x}_1 \in \mathcal{X}$ dominates another solution $\mathbf{x}_2 \in \mathcal{X}$ (denoted with $\mathbf{x}_1 \prec \mathbf{x}_2$), iff[1]

$$\forall i \in \{1, ..., d\} : f_i(\mathbf{x}_1) \leq f_i(\mathbf{x}_2) \text{ and}$$
$$\exists k \in \{1, ..., d\} : f_k(\mathbf{x}_1) < f_k(\mathbf{x}_2). \tag{1}$$

Solutions $\mathbf{x}_1$ and $\mathbf{x}_2$ are incomparable, when neither $\mathbf{x}_1 \prec \mathbf{x}_2$ nor $\mathbf{x}_2 \prec \mathbf{x}_1$. Incomparable solutions, which are not dominated by any other solution found so far, are called the non-dominated front.

---

[1] For simplicity, we describe only the minimisation of objective functions, since maximisation can be achieved by minimising the function with its sign reversed.

When a multi-objective optimisation algorithm outputs the non-dominated front $\mathbf{x}_1, ..., \mathbf{x}_N$, it may be still required to evaluate solutions individually, for example, to select better solutions after evolutionary mutation. This can be done by means of the *dominated hypervolume*, or *S*-metric [34], which is estimated as follows. Let $\mathbf{r} \in \mathcal{F}$ be a reference point in the objective space, which corresponds to the worst possible solution (e.g., 1 for the dimension which corresponds to the classification error). Then we define the dominated hypervolume as:

$$H(\mathbf{x}_1, ..., \mathbf{x}_N) = vol \left(\bigcup_{i=1}^{N} [\mathbf{x}_i, \mathbf{r}] \right), \tag{2}$$

where $vol(\cdot)$ describes the volume in $\mathbb{R}^d$, and $[\mathbf{x}_i, \mathbf{r}]$ the hypercube spanned between $\mathbf{x}_i$ and $\mathbf{r}$. Generally speaking, the dominated hypervolume of a given front consists of an infinite number of all theoretically possible solutions which are always dominated by solutions in this front. An individual hypervolume contribution of a solution $\mathbf{x}_i$ is then estimated as the dominated hypervolume of the front without this solution:

$$\Delta H(\mathbf{x}_i) = H(\mathbf{x}_1, .., \mathbf{x}_N) - H(\mathbf{x}_1, ..\mathbf{x}_{i-1}, \mathbf{x}_{i+1}, .., \mathbf{x}_N). \tag{3}$$

S-metric selection evolutionary multi-objective algorithm (SMS-EMOA) [3] generates exactly one solution in each iteration. Then, all solutions are assigned to a hierarchy of non-dominated fronts, and the solution from the worst front with the smallest $\Delta H(\mathbf{x}_i)$ is removed from the population. For more details, we refer to [3].

3.2   Application to Training Set Selection

In this study, we apply SMS-EMOA to evolutionary multi-objective training set selection (EMO-TSS): We simultaneously minimise (a) the ratio $f_{ir} := \frac{N}{I}$ between the number of selected instances N from the training set and its total number of instances I and (b) the balanced classification error f_e achieved with a classification model trained on the given selection of instances which is computed on a separate validation set as the average of misclassification rates for positive and negative instances in binary classification. Each TSS solution is represented by a binary vector of length I called representation vector, where a one at the i-th position means that the i-th instance is used to train the classification model. Note this also supports optimising the selection of data augmentations when they are included as additional instances in the training set.

In the beginning of the evolutionary loop, the instances are selected for each individual with a given probability p. Using all instances ($p = 1$) would decrease the diversity of the initial population and thus increase the optimisation time. Additionally, selecting too few instances for each individual would increase the danger that "smart" instances not initially contained in the population are never found, so we set $p = 0.5$. To encourage the search for solutions with few training instances, we set a higher probability to remove ones than to add ones during mutation. For each position i of a new offspring solution, the i-th bit is flipped from 1 to 0 with probability w_1/N. The bit is flipped from 0 to 1 with probability $w_2 \cdot (w_1/N)$. We set these parameters so that solutions with only few training instances are explored more thoroughly (see Sect. 4.3).

4   Experimental Setup

In the following, we experimentally validate our proposed multi-objective training data selection algorithm for music classification tasks for the example of vocal activity detection.

4.1   Data Sets

We have selected MedleyDB [4] as a data set for vocal detection, as it contains multitrack audio recordings allowing for the application of individual data augmentation strategies for vocals and accompaniment. From the original 122 multi-track recordings, we removed 8 tracks due to source interference in the recordings[2].

To accelerate feature extraction and classification, we extracted 10 snippets[3] of 3 s duration from each track, equally distributed along the track length. Based on the instrument activation labels, three variants of each snippet were created that contain only vocals, only accompaniment, and a mix of both in case vocals and accompaniment were available, respectively. For each accompaniment snippet, we additionally created 4 versions whose signal amplitude was reduced to 80%, 60%, 40% and 20% and 8 further variants by applying the following degradations from the Audio Degradation Toolbox [16]: live recording, strong compression, vinyl recording, four noises with signal-to-noise ratio (SNR) of 20 dB (pink, blue, violet, white), and adding a sound 'OldDustyRecording' with a SNR of 10 dB. For vocal snippets, we also created 4 quieter, but no degraded versions, because we restricted the focus of this study to recognise "clean" vocals along varied accompaniment sounds. For audio snippets from time intervals which contained both accompaniment and vocals, we created mixes of the vocal snippet with the original accompaniment snippet and all augmented variants of accompaniment snippets. Together with non-mixed vocal and accompaniment snippets, this strategy leads to an overall number of 22,953 snippets with annotations. Naturally, the search space for EMO-TSS can be almost arbitrarily extended using further augmentations, but the above augmentations serve to demonstrate the potential of our approach.

---

[2] *Grants-PunchDrunk* contained vocals in the "sampler" stem, and for the following tracks, vocals could be heard in non-vocal stems: *ClaraBerryAndWooldog-TheBadGuys, Debussy-LenfantProdigue, Handel-TornamiAVagheggiar, Mozart-BesterJungling, Mozart-DiesBildnis, Schubert-Erstarrung, Schumann-Mignon*.

[3] As data *instances* are represented by audio *snippets* in our study, we use both terms synonymously throughout this paper.

In our experimental setup, we distinguish between three data sets. A *training* set is used by EMO-TSS to create a representation vector (see Sect. 3.1) and to choose the snippets used for training the vocal activity classifier. A *validation* set is used to evaluate solutions (trained classification models) created by EMO-TSS. A *test* set is used for the independent evaluation of the best solutions found after the evolutionary optimisation. From 114 MedleyDB tracks, one fourth was used as a training set, one half as a validation set, and one fourth as a test set. This partitioning was done three times similarly to a 3-fold cross-validation approach to estimate the variability of results depending on dataset selection.

As the goal of our training set selection was to identify the most relevant snippets and their augmentations for the detection of vocals in "usual" non-augmented recordings, augmented snippets were used only in the training set and not in validation and test sets. To evaluate the impact of the set of automatically selected data augmentations, we introduce two selections as a baseline for comparison: the first one ("pure snippets") uses all original snippets of the training set without augmentations, and the second one ("all snippets") uses all snippets including their augmented variants.

4.2   Features

As a first step before feature extraction, we convert the input audio to mono sampled at 22050 Hz.

As the first feature set, we use a vector of 13 Mel Frequency Cepstral Coefficients (MFCCs) [23], which were developed for speech recognition and should therefore facilitate vocal detection well. These features were computed individually for the middle of the attack interval, onset frame, and the middle of the release interval (extracted with an onset detection function from MIR Toolbox [12]), leading to a 39-dimensional feature vector for each snippet. We distinguish between these three time periods, because instrument and vocal timbre may significantly change during the time a note is played; for example, in [14] it was shown that non-harmonic properties in the attack phase may be very helpful to identify instruments, and the frames with a stable sound between two onsets performed well for genre recognition in [31].

As the second feature set, we use a Log-Mel-Spectrogram (LMS). We compute the magnitude spectrogram resulting from a 1024-point FFT with a hop size of 512, which is then converted to a Mel-frequency spectrogram using a Mel filterbank with 40 bands ranging from 60 to 8000 Hz. Finally, we normalise the features x by applying log-normalisation: $x \rightarrow \log(1 + x)$. To reduce feature dimensionality, we then downsample the spectrogram using mean-pooling across time frames with a non-overlapping window of size four, after discarding time frames at borders so that the number of time frames is divisible by four. The result is a $F \times T$ feature matrix with $F = 40$ frequencies and $T = 32$ time frames for a total of 1280 features.

We omitted the usage of larger audio feature sets, since they could lead to stronger overfitting to the validation set, especially when exploring extremely sparse solutions that use only a few instances for training. However, given a sufficiently large validation set and training time, we expect our method to work in these cases as well. Furthermore, instances important for vocal recognition performance should still be preferred over others rather independently of the feature set used.

4.3   Classification Models and Training Set Selection

We applied a random forest (RF) [5] as our classification method, as it is fast, has only few parameters, and is robust to overfitting, which keeps the meta-optimisation feasible with regards to computing time. For the same reasons, we refrained from the use of deep neural networks in this study.

However, we investigated to which extent the optimised training set selection obtained when using the simple RF classifier can also improve classification performance when it is used to train more complex classifiers. For this, we first optimised the training set selection using the RF classifier, serving as a fast surrogate model for a larger neural network we reimplemented from previous work for vocal detection [27]. This network is trained with the same settings on all, pure and the optimised EMO-TSS training set obtained using the RF, using the downsampled Log-Mel-Spectrogram features from Sect. 4.2.

However, we did not achieve improvements in classification performance compared to simply using pure snippets. We hypothesise that this is because of the drastically different approach neural networks employ for classification compared to RF, and that the EMO-TSS optimisation adapts closely to the behaviour of the used classification model. However, note that, in principle, our EMO-TSS approach can still be used to potentially improve the performance of this neural network by using it directly as part of the evolutionary meta-optimisation, although this significantly increases the required computing time. We leave investigations with larger models and suitable smaller surrogate models that behave sufficiently similar during evolutionary optimisation for future work.

For SMS-EMOA, we found the following parameters to work well after a preliminary study: initialisation with approximately one half of randomly selected instances, a population size of 40, $w_1 = 0.05$, and $w_2 = 0.2$. The number of generations was set to 3000, and each experiment was repeated 15 times for each fold. Please note that an exhaustive search for the optimal settings was not the target of this study and could be addressed in future work.

5   Results and Discussion

We analyse the results of training selection in Sect. 5.1 and investigate the properties of the chosen instances and augmentation strategies in Sect. 5.2.

5.1   Number of Selected Snippets and Classification Performance

Figure 1 plots the optimisation progress for the first cross-validation fold, averaged across 15 statistical experiment repetitions. In Fig. 1(a), we observe a strong increase in hypervolume over 3000 generations for the validation set. This is due to starting with a rather "poor" solution with approximately half of all instances, so that there is enough room for improvement by removing many training instances. In Fig. 1(b), the hypervolume for the test set also strongly increases, but remains below the hypervolume for the validation set because of higher test classification errors. In Fig. 1(c), the progress of the lowest error $\widehat{f_e}$ among non-dominated solutions for the validation set is plotted, again averaged across 15 repetitions. Similar to the change of hypervolume with an increasing evaluation number, the optimisation progress is faster at the beginning and slows down significantly around the second third of evaluations. Figure 1(d) presents the progress of $\widehat{f_e}$ for the test set. As expected, the test error does not fall as rapidly as the validation error; the difference in values for the first generation is explained by a varying distribution of music tracks between the validation and test sets. Importantly, we do not observe significant overfitting on the validation set, which would be visible as an increasing test error.

Figure 2 shows the non-dominated front after all EMO-TSS experiments for fold 1. For the validation set, the solutions are marked with rectangles. Diamonds correspond to the same solutions evaluated for the test set: the ratio of selected instances remains the same, but the classification error increases slightly due to overfitting to the validation set. This increase is much more variable for smaller ratios, because the model parameters can then vary more strongly depending on the particular training set selection. Note that we assign a classification error of 1 to extremely sparse training set selections with only vocal or only non-vocal snippets and do not show them here. The solution with the smallest validation error $f_e = 0.2055$ contains 11.73% of all snippets (714 snippets), but when we further reduce the number of selected snippets to 434 snippets, f_e increases only moderately to 0.2125, indicating diminishing returns when adding more snippets and an effective optimisation on the validation set.

Table 1 lists the results after optimisation. To measure the effect of EMO-TSS, we compare the results to two baselines "pure snippets" and "all snippets", cf. Sect. 4.1. With "all snippets", the ratio of selected instances $\widehat{f_{ir}} = 1$, and with "pure snippets" $\widehat{f_{ir}}$ is equal to the number of non-augmented training snippets divided by the number of all snippets (including augmented ones). The complete training set with all augmentations contains 6088 snippets for fold 1, 5477 snippets for fold 2, and 5494 snippets for fold 3. We make several observations in the following.

OBSERVATION 1–COMPARISON OF BASELINES: Extension of pure snippets with all augmentations leads to an increase of the error in all cases: for both validation and test sets and both feature vectors. In other words, uncontrolled extension of the training set does not lead to an increase of classification performance, but even to its reduction. That not every augmentation is equally useful was also noted in work on manually selected augmentation strategies for singing voice detection [27] and demonstrates the data augmentation is not always straightforward to apply correctly.

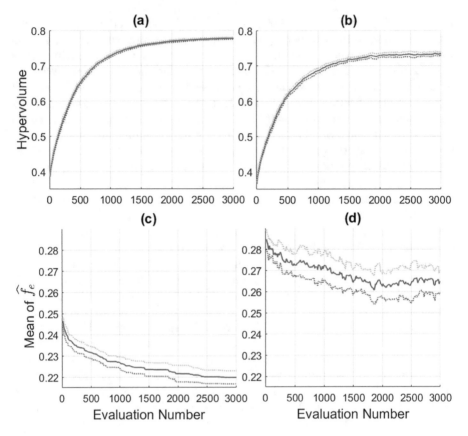

Fig. 1. Optimisation progress and 95% confidence intervals over 3000 evaluations for all statistical experiment repetitions: mean dominated hypervolume for the validation set (a) and the test set (b); mean of the smallest achieved error on the validation set (c) and the test set (d).

OBSERVATION 2–EMO-TSS REDUCES CLASSIFICATION ERROR OR THE NUMBER OF SELECTED INSTANCES: For all folds and features, our approach either improves the test performance $\hat{f}_e(T)$ compared to baseline methods (e.g., $\hat{f}_e(T) = 0.2667$ for the 1st fold using MFCCs, smaller than 0.2848 with pure snippets and 0.3489 with all snippets), or reaches similar performance, but using 39.5% less instances than the "Pure" baseline on average. This supports our initial hypothesis that many training instances can be unrepresentative or do not provide much new information to the classifier. A reduction in training instances can be beneficial for gaining an understanding of the classification problem and for example when using non-parametric models (e.g. k-nearest neighbour) whose storage space grows with the amount of training data.

OBSERVATION 3–COMPARISON OF VALIDATION ERRORS: Using our proposed method, both the number of selected instances and the validation error is

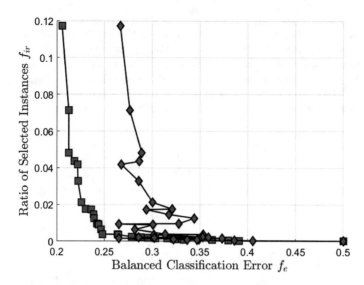

Fig. 2. Non-dominated front for the first fold, on the validation set (rectangles) and on the test set (diamonds).

consistently reduced across all features and folds compared to the baseline methods, e.g., from $\{0.2411, 0.2843\}$ to 0.2055 for fold 1 using MFCCs and from $\{0.1572,$ $0.1916\}$ to 0.1334 using LMS. This means that EMO-TSS is very effective at optimising the training set selection to achieve good validation performances.

OBSERVATION 4–COMPARISON OF EMO-TSS VALIDATION AND TEST ERRORS: When applying TSS, the most critical risk is that its selection might lead to a model with good validation performance, but worse test set performance due to overfitting. Indeed, test errors sometimes increase slightly in comparison, but always remain approximately as good as the baseline test errors. To further reduce overfitting towards the validation set in the future, we suggest using a larger and more representative validation set by integrating multiple databases with vocal annotations.

OBSERVATION 5–COMPARISON OF $\widetilde{f}_e(T)$ AFTER EMO-TSS to baselines: From all non-dominated solutions returned by EMO-TSS for an individual fold, we identified the one with the smallest test error $\widetilde{f}_e(T)$. Among all combinations of folds and features, this error is lower than the test error $\widetilde{f}_e(T)$ of the best baseline in almost all cases. This means that it is possible to obtain reduced test errors also after long-iterated TSS for the validation set. However, it is not possible to directly identify this "oracle" solution when the annotations of the test set are unseen, because the solution with the smallest test error is not the same as the solution with the smallest validation error, so this is a more theoretical result.

OBSERVATION 6–COMPARISON OF FOLDS: The errors are fairly different across all folds, for instance, the smallest test errors are achieved for fold 3. This is explained by a rather small size of MedleyDB: test sets contained snippets of 28 tracks (one fourth of 114 tracks).

Table 1. Impact of our approach on classification performance. F corresponds to the fold, $\widehat{f}_{ir}$ to the ratio of selected instances for the non-dominated solution with the smallest classification error across all statistical repetitions, $\widehat{f}_e(V)$ and $\widehat{f}_e(T)$ to the classification error of the same solution on the validation and test set, and $\widetilde{f}_e(T)$ denotes the smallest test error across all final non-dominated solutions. Note that $\widehat{f}_e(T) = \widetilde{f}_e(T)$ for baseline methods "Pure" and "All" without training set selection. The best validation and test errors for each fold are marked with bold font.

F	Features	Snippets	$\widehat{f}_{ir}$	$\widehat{f}_e(V)$	$\widehat{f}_e(T)$	$\widetilde{f}_e(T)$
1	MFCCs	Pure	0.0907	0.2411	0.2848	0.2848
		All	1.0000	0.2843	0.3489	0.3489
		EMO-TSS	0.1173	0.2055	0.2667	0.2433
	LMS	Pure	0.0907	0.1572	0.1322	0.1322
		All	1.0000	0.1916	0.1903	0.1903
		EMO-TSS	0.0866	**0.1334**	0.1679	**0.1281**
2	MFCCs	Pure	0.0893	0.2752	0.3265	0.3265
		All	1.0000	0.2867	0.3297	0.3297
		EMO-TSS	0.0416	0.2127	0.3166	0.2959
	LMS	Pure	0.0893	0.1860	**0.1971**	**0.1971**
		All	1.0000	0.2609	0.3028	0.3028
		EMO-TSS	0.0310	**0.1339**	0.2199	0.2023
3	MFCCs	Pure	0.0885	0.2937	0.2592	0.2592
		All	1.0000	0.3068	0.2782	0.2782
		EMO-TSS	0.0220	0.2506	0.2700	0.2265
	LMS	Pure	0.0885	0.1642	0.1800	0.1800
		All	1.0000	0.2495	0.2049	0.2049
		EMO-TSS	0.0284	**0.1432**	0.1801	**0.1562**

5.2 Properties of Selected Instances

A further analysis may be done with regards to categories of selected snippets in solutions with the smallest classification errors. Can we recommend applying some particular augmentations generally, not just to specific instances? For this we use a measure describing the "popularity" of a set of snippets, obtained by dividing the frequency of their occurrence in the optimised training set selection by the frequency of their occurrence in the full training set. Table 2 shows this measure for different snippet categories computed for solutions with the smallest f_e. For vocal snippets on fold 1 with MFCCs as an example, this yields $1.028 \approx \frac{0.1149}{0.1117}$, because the proportion of vocal snippets is 0.1117 in the complete training set (680 of 6088 snippets) and 0.1149 for the training set selection with the smallest $f_e(V)$ in the first fold (82 out of 714 snippets). Numbers greater than 1 show that the proportion of a given snippet category is larger in the optimised set than in the complete set, indicating its importance for high performance. Numbers lower than 1 suggest a lower importance.

Table 2. Share of different snippet categories in relation to the share of these snippets in the complete training set for the three folds.

Snippet category	MFCCs			LMS		
	$F = 1$	$F = 2$	$F = 3$	$F = 1$	$F = 2$	$F = 3$
Main categories						
Vocals	1.028	1.179	1.234	1.087	1.640	1.208
Accompaniment	0.998	0.907	0.836	0.930	0.773	0.826
Mix	0.994	1.183	1.356	1.111	1.363	1.394
Applied augmentations						
Ampl. reduct. 80%	1.081	1.081	1.215	0.879	0.922	0.951
Ampl. reduct. 60%	1.128	0.884	0.747	0.858	0.857	1.207
Ampl. reduct. 40%	0.958	1.130	1.215	1.130	1.384	0.878
Ampl. reduct. 20%	1.020	1.032	0.654	1.109	0.857	1.427
Live	1.004	0.634	0.830	0.889	0.935	0.789
Compression	0.963	0.951	0.356	0.972	1.360	1.439
Vinyl recording	0.922	1.078	1.067	0.972	0.850	0.975
Noise: pink	1.127	0.761	1.186	0.916	0.680	0.418
Noise: blue	1.086	0.761	1.067	1.055	0.935	1.021
Noise: violet	0.963	1.141	1.067	0.916	0.340	0.882
Noise: white	0.840	0.824	0.474	0.972	1.615	0.696
Old dusty	0.943	1.014	1.423	1.277	0.850	1.300

Results show that vocal snippets are rather relevant, with their measures above 1 for all folds and both features. This reveals it is helpful to train vocal detection models not only with music mixtures, but also solo vocal tracks as positive examples. However, the numbers are not significantly higher than 1 for some folds. Generally, no category of snippets appears to be very important or very harmful to include; all categories contribute to models with the lowest error to some extent. Interestingly, the effect of augmentations seems to be somewhat dependent on the used features: for MFCCs, the most valuable augmentation seems to be the reduction of signal amplitude to 80% (ratios greater than 1 for all folds), while for LMS it appears harmful (ratios below 1 for all folds). Some values vary strongly across the folds. For instance, white noise degradation is less important for all folds using MFCCs and two of three LMS folds, but has a rather high value of 1.615 for the second LMS fold. Further investigations with larger data sets and also tracks from different music genres are necessary to potentially provide more conclusive findings about the relevance of individual augmentation methods.

6   Conclusions

In this work, we have proposed a multi-objective evolutionary framework for training set selection, which simultaneously minimises the number of training instances and the classification error. This approach was applied for the problem of vocal detection, together with training set augmentation by means of loudness reduction and various audio degradations. The results show that, compared to classification with a complete training set, it is possible to strongly reduce the training set size while approximately maintaining the classification performance. Using our optimised selection of training instances and augmentations, we obtain a strong performance increase on the validation set compared to using all or no augmentations for training, which mostly translates to an independent test set, albeit sometimes to a lesser degree due to over-optimisation on the validation set.

To improve the generalisation performance and make validation sets more representative, one can integrate further databases with vocal annotations in our proposed framework. It would be interesting to explore the performance of our proposed training set selection when using more feature sets, classification methods and augmentation methods and other parameters for the tested augmentation methods, and even applying it to other tasks in music information retrieval. Also, we may compare the performance of evolutionary optimisation to other training set selection techniques, like clustering or n-gram statistics [30].

Finally, for the application to larger classification models such as neural networks, it appears promising to investigate the use of surrogate classifier models that are fast to train and behave similarly to the large model of interest, so that the computation time for evolutionary optimisation remains feasible, and the optimal training set selection found for the surrogate model also helps the performance for the large classification model.

Acknowledgements. This work was funded by the DFG (German Research Foundation, project 336599081) and by EPSRC grant EP/L01632X/1.

References

1. Acampora, G., Herrera, F., Tortora, G., Vitiello, A.: A multi-objective evolutionary approach to training set selection for support vector machine. Knowl. Based Syst. **147**, 94–108 (2018)
2. Bäck, T.: Evolutionary Algorithms in Theory and Practice. Oxford University Press, New York (1996)
3. Beume, N., Naujoks, B., Emmerich, M.: SMS-EMOA: multiobjective selection based on dominated hypervolume. Eur. J. Oper. Res. **181**(3), 1653–1669 (2007)
4. Bittner, R.M., Salamon, J., Tierney, M., Mauch, M., Cannam, C., Bello, J.P.: MedleyDB: a multitrack dataset for annotation-intensive MIR research. In: Proceedings of the 15th International Society for Music Information Retrieval Conference (ISMIR), pp. 155–160 (2014)
5. Breiman, L.: Random forests. Mach. Learn. **45**(1), 5–32 (2001)

6. Cano, J.R., Herrera, F., Lozano, M.: Evolutionary stratified training set selection for extracting classification rules with trade off precision-interpretability. Data Knowl. Eng. **60**(1), 90–108 (2007)
7. Coello, C.A.C., Lamont, G.B., Veldhuizen, D.A.V.: Evolutionary Algorithms for Solving Multi-Objective Problems. Springer, New York (2007). https://doi.org/10.1007/978-0-387-36797-2
8. Defferrard, M., Benzi, K., Vandergheynst, P., Bresson, X.: FMA: a dataset for music analysis. In: Proceedings of the 18th International Society for Music Information Retrieval Conference (ISMIR), pp. 316–323 (2017)
9. Fujinaga, I.: Machine recognition of timbre using steady-state tone of acoustic musical instruments. In: Proceedings of the International Computer Music Conference (ICMC), pp. 207–210 (1998)
10. Goto, M., Nishimura, T.: RWC music database: popular, classical, and jazz music databases. In: Proceedings of the 3rd International Conference on Music Information Retrieval (ISMIR), pp. 287–288 (2002)
11. Kumar, A., Cowen, L.: Augmented training of hidden Markov models to recognize remote homologs via simulated evolution. Bioinformatics **25**(13), 1602–1608 (2009)
12. Lartillot, O., Toiviainen, P.: MIR in Matlab (II): a toolbox for musical feature extraction from audio. In: Proceedings of the 8th International Conference on Music Information Retrieval (ISMIR), pp. 127–130 (2007)
13. Lemley, J., Bazrafkan, S., Corcoran, P.: Smart augmentation learning an optimal data augmentation strategy. IEEE Access **5**, 5858–5869 (2017)
14. Livshin, A., Rodet, X.: The significance of the non-harmonic "noise" versus the harmonic series for musical instrument recognition. In: Proceedings of the 7th International Conference on Music Information Retrieval (ISMIR), pp. 95–100 (2006)
15. Mäkinen, T., Kiranyaz, S., Pulkkinen, J., Gabbouj, M.: Evolutionary feature generation for content-based audio classification and retrieval. In: Proceedings of the 20th European Signal Processing Conference (EUSIPCO), pp. 1474–1478 (2012)
16. Mauch, M., Ewert, S.: The audio degradation toolbox and its application to robustness evaluation. In: Proceedings of the 14th International Society for Music Information Retrieval Conference (ISMIR), pp. 83–88 (2013)
17. Mauch, M., Fujihara, H., Yoshii, K., Goto, M.: Timbre and melody features for the recognition of vocal activity and instrumental solos in polyphonic music. In: Proceedings of the 12th International Society for Music Information Retrieval Conference (ISMIR), pp. 233–238 (2011)
18. McFee, B., Humphrey, E.J., Bello, J.P.: A software framework for musical data augmentation. In: Proceedings of the 16th International Society for Music Information Retrieval Conference (ISMIR), pp. 248–254 (2015)
19. Mierswa, I., Morik, K.: Automatic feature extraction for classifying audio data. Mach. Learn. J. **58**(2–3), 127–149 (2005)
20. Miranda, E.R., Biles, J.A.: Evolutionary Computer Music. Springer, New York (2007). https://doi.org/10.1007/978-1-84628-600-1
21. Mun, S., Park, S., Han, D.K., Ko, H.: Generative adversarial network based acoustic scene training set augmentation and selection using SVM hyper-plane. In: Proceedings of the Detection and Classification of Acoustic Scenes and Events 2017 Workshop (DCASE 2017), November 2017
22. Pachet, F., Zils, A.: Evolving automatically high-level music descriptors from acoustic signals. In: Wiil, U.K. (ed.) CMMR 2003. LNCS, vol. 2771, pp. 42–53. Springer, Heidelberg (2004). https://doi.org/10.1007/978-3-540-39900-1_5
23. Rabiner, L., Juang, B.H.: Fundamentals of Speech Recognition. Prentice Hall, Upper Saddle River (1993)

24. Rao, V., Gupta, C., Rao, P.: Context-Aware features for singing voice detection in polyphonic music. In: Detyniecki, M., García-Serrano, A., Nürnberger, A., Stober, S. (eds.) AMR 2011. LNCS, vol. 7836, pp. 43–57. Springer, Heidelberg (2013). https://doi.org/10.1007/978-3-642-37425-8_4

25. Regnier, L., Peeters, G.: Singing voice detection in music tracks using direct voice vibrato detection. In: Proceedings of the IEEE International Conference on Acoustics, Speech and Signal Processing (ICASSP), pp. 1685–1688. IEEE (2009)

26. Schlüter, J.: Learning to pinpoint singing voice from weakly labeled examples. In: Proceedings of the International Society for Music Information Retrieval Conference (ISMIR), pp. 44–50 (2016)

27. Schlüter, J., Grill, T.: Exploring data augmentation for improved singing voice detection with neural networks. In: Proceedings of the 16th International Society for Music Information Retrieval Conference (ISMIR), pp. 121–126 (2015)

28. Stoller, D., Ewert, S., Dixon, S.: Jointly detecting and separating singing voice: a multi-task approach. In: Deville, Y., Gannot, S., Mason, R., Plumbley, M.D., Ward, D. (eds.) LVA/ICA 2018. LNCS, vol. 10891, pp. 329–339. Springer, Cham (2018). https://doi.org/10.1007/978-3-319-93764-9_31

29. Vatolkin, I., Preuß, M., Rudolph, G.: Multi-objective feature selection in music genre and style recognition tasks. In: Proceedings of the 13th Annual Genetic and Evolutionary Computation Conference (GECCO), pp. 411–418. ACM Press (2011)

30. Vatolkin, I., Preuß, M., Rudolph, G.: Training set reduction based on 2-gram feature statistics for music genre recognition. Technical report TR13-2-001, Faculty of Computer Science, Technische Universität Dortmund (2013)

31. Vatolkin, I., Theimer, W., Botteck, M.: Partition based feature processing for improved music classification. In: Gaul, W.A., Geyer-Schulz, A., Schmidt-Thieme, L., Kunze, J. (eds.) Challenges at the Interface of Data Analysis, Computer Science, and Optimization. Studies in Classification, Data Analysis, and Knowledge Organization, pp. 411–419. Springer, Heidelberg (2012). https://doi.org/10.1007/978-3-642-24466-7_42

32. Velasco, J.M., et al.: Data augmentation and evolutionary algorithms to improve the prediction of blood glucose levels in scarcity of training data. In: Proceedings of the 2017 IEEE Congress on Evolutionary Computation (CEC), pp. 2193–2200. IEEE (2017)

33. Zitzler, E.: Evolutionary multiobjective optimization. In: Rozenberg, G., Bäck, T., Kok, J.N. (eds.) Handbook of Natural Computing, vol. 2, pp. 871–904. Springer, Heidelberg (2012)

34. Zitzler, E., Thiele, L.: Multiobjective optimization using evolutionary algorithms — a comparative case study. In: Eiben, A.E., Bäck, T., Schoenauer, M., Schwefel, H.-P. (eds.) PPSN 1998. LNCS, vol. 1498, pp. 292–301. Springer, Heidelberg (1998). https://doi.org/10.1007/BFb0056872

Automatic Jazz Melody Composition Through a Learning-Based Genetic Algorithm

Yong-Wook Nam and Yong-Hyuk Kim$^{(\boxtimes)}$

Department of Computer Science, Kwangwoon University, Seoul, Republic of Korea
{mitssi,yhdfly}@kw.ac.kr

Abstract. In this study, we automate the production of good-quality jazz melodies through genetic algorithm and pattern learning by preserving the musically important properties. Unlike previous automatic composition studies that use fixed-length chromosomes to express a bar in a score, we use a variable-length chromosome and geometric crossover to accommodate the variable length. Pattern learning uses the musical instrument digital interface data containing the jazz melody; a user can additionally learn about the melody pattern by scoring the generated melody. The pattern of the music is stored in a chord table that contains the harmonic elements of the melody. In addition, a sequence table preserves the flow and rhythmic elements. In the evaluation function, the two tables are used to calculate the fitness of a given excerpt. We use this estimated fitness and geometric crossover to improve the music until users are satisfied. Through this, we successfully create a jazz melody as per user preference and training data.

Keywords: Genetic algorithm · Automatic composing · Geometric crossover

1 Introduction

Since a long time, humans have used automation to maximize productivity. In the industrial age, simple labor is designed to be performed almost entirely by a machine instead of a human. In the information age, computers can automatically judge and process tasks like humans. The ability of thinking like humans is called artificial intelligence (AI). AI has become very important that there is no place that does not have an impact in our lives. Even now, human actions are being recorded as data that will be used for AI to make better judgments. This is also true of compositions in art. There has been a lot of effort to use AI to compose music like humans. Before deep learning became popular, there had been many attempts to implement automatic composition through evolutionary computation (EC). In particular, genetic algorithms (GAs), which are a major part of EC, have been widely used. The process of creating good music

© Springer Nature Switzerland AG 2019
A. Ekárt et al. (Eds.): EvoMUSART 2019, LNCS 11453, pp. 217–233, 2019.
https://doi.org/10.1007/978-3-030-16667-0_15

using a GA is not much different from the process of solving other optimization problems. By selecting good chromosomes in the population, new offspring are created through crossover; this process is repeated to produce a better chromosome. Usually, most optimization problems focus on creating an offspring with a higher fitness score more efficiently based on a well-defined evaluation function. However, it is very difficult to evaluate the fitness in automatic composition research. For a human to evaluate a certain music, his/her taste, musical knowledge, experience, environment, and culture may be collectively used [1, 2]. Therefore, designing a fitness function is still one of the most challenging problems in automatic composition research. In this study, we attempt to solve this problem by training not only the harmonics but also patterns of a music score. Hence, if the music to be evaluated is similar to the learned pattern, the evaluation function will assign it a high score. Through this, the music produced will be similar to the music learned.

The most important aspect in GAs is not just automatic composition but the crossover operator. Even if good chromosomes are chosen, it is difficult to achieve good performance unless they pass their good qualities onto their offspring. The representation of the chromosome has a significant effect on the crossover operator. Choosing a note as a gene is desirable because it is aimed to create a good melody. However, because a note is time-series data with a duration, the number of notes in a bar is variable. Therefore, many crossover techniques cannot be used because the parent lengths are not the same. Therefore, previous studies converted the variable-length chromosome into a fixed-length chromosome [3–5]. In this study, we use a crossover operator for variable-length chromosomes instead. This results in the evolution of offspring while preserving the good traits of the parents.

Music is as diverse as its history. People's favorite music changes according to their region and age and also the trend. In that sense, jazz is a appropriate genre for automatic composition research, which is generally recognized as "good" by people; therefore, in this study, we choose jazz music as well. Musical instrument players play jazz with a swing rhythm. This rhythm is not usually referred to as a triplet note. However, for convenience, we use a triplet note that is similar to swing rhythm. Our contribution to the automatic composition research is summarized as follows:

(i) We create a model that learns patterns of musical notes; the model can evaluate music based on patterns of learned music.
(ii) We successfully recombine variable-length note chromosomes using a novel geometric crossover to successfully transfer genes with good traits.

This paper is organized as follows: Sect. 2 describes related technologies and studies; Sect. 3 explains the learning score tables, a model to determine fitness; Sect. 4 explains the GA elements used in this study; Sect. 5 describes the experiments; and we draw conclusions in Sect. 6.

2   Literature Review

Honer et al. [6] published a theory that stated that algorithmic composition is possible using a GA. Biles [4] succeeded in creating a jazz melody using a GA. Based on this research, Biles made a program named "Genjam" and showed the possibility of collaborative performance between humans and AI. Quick [7] developed a Haskell-based machine learning framework called Kulitta, which demonstrated that it was possible to produce musical products similar to classical music masterpieces, especially the ones written by Bach. PG-Music's Band-in-a-Box[1] is a music program that provides computer-generated background music for a musician. This program also has an algorithm that creates a solo melody. This algorithm is based on an expert system, but produces a high quality melody. Impro-Visor is a program that helps users create jazz solos [8]. There are several studies [9,10] that use this program to make good jazz melody. Fernández de Vega [11] presented GA-based techniques that evolve chord progressions. He showed that 4-part harmonization can also be created using GAs. We have previously published a short paper on automatic composition using geometric crossover and pattern learning [12,13]. However, the genre was vocal music, and there was a problem of similar melody generation.

There has been much recent active research on automatic composition using deep learning. Johnson's research [14] produced music using recurrent neural networks (RNNs), and piano music created by emulating Chopin's style. However, the RNN was often overfit and sometimes played scales that cannot be understood harmonically. Hadjeres and Pachet [15] also used an RNN and emulated Bach's style to create quartet music. The automatic composition project "*FlowMachines*"[2] showcased *Daddy's Car*, created by copying Beatles' style.

3   Learning Note Patterns and Evaluation

This section describes how to learn and evaluate music. Usually, in jazz music, surrounding instruments play a chord and the solo instrument plays a solo melody. If this melody does not fit the chord, there is a lack of harmony, which could result in bad-quality music. Therefore, this study used machine learning to determine which notes fit a chord according to the theory of harmony and how notes can complement each other; we designed a program to compute the fitness using an established learning model. We used two learning models: (1) a chord score table, which is a model that determines the fitness according to the theory of harmony; and (2) a sequence score table, which is a model that determines whether the sequence of notes is good or not. When a user gives a rating to a musical score with chords and melodies prepared for learning, the two score tables are updated using information obtained from the music. Because every score table has different rating criteria and a particular element could have a high score depending on the criteria, the scores must be normalized when computing

---

[1] http://www.pgmusic.com/about.htm.

[2] http://www.flow-machines.com.

the fitness. In this study, scores were normalized by dividing all elements so that the highest rating was 1 in the score table. Bad music can get a negative score, in which case the elements of those notes will have their scores removed. Even if the actual score is negative, however, it was converted to 0 when normalized.

3.1 Chord Score

The chord score table numerically shows how fit a certain pitch is for a particular chord according to the theory of harmony. Each note has its own pitch, and a chord played in the background is a combination of these pitches. Figure 1 shows a chord score table that demonstrates the relationship between chords and scales. This shows the initialized CMaj7 chord score table, where no change is made yet through learning. Column 0 '@' is a one-note score table and shows the score of a single note in the CMaj7 chord. Columns 1 to 12 are two-note score tables that contain scores for when a note in the column is played next (e.g., the score in Row 3 and Column 2 indicates a score when Note C and Note D are played next). Because a melody is a sequence of consecutive notes, the information will be more meaningful when the pattern of the next note is learned.

Scores in the initialized score table were assigned in accordance with the chord scale. Musical tonality was fixed as C major and fitted to a scale suitable for jazz music. Here, "tension" and "avoid note" were not considered. This is because although the notes are not contained in the chord, they can be learned from the training data. If they are used in learning data in the future, a good score should be assigned to these notes.

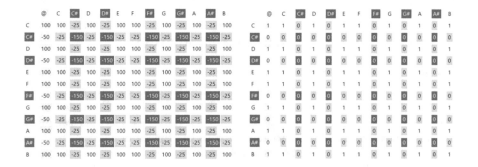

	@	C	C#	D	D#	E	F	F#	G	G#	A	A#	B
C	100	100	-25	100	-25	100	100	-25	100	-25	100	-25	100
C#	-50	-25	-150	-25	-150	-25	-25	-150	-25	-150	-25	-150	-25
D	100	100	-25	100	-25	100	100	-25	100	-25	100	-25	100
D#	-50	-25	-150	-25	-150	-25	-25	-150	-25	-150	-25	-150	-25
E	100	100	-25	100	-25	100	100	-25	100	-25	100	-25	100
F	100	100	-25	100	-25	100	100	-25	100	-25	100	-25	100
F#	-50	-25	-150	-25	-150	-25	-25	-150	-25	-150	-25	-150	-25
G	100	100	-25	100	-25	100	100	-25	100	-25	100	-25	100
G#	-50	-25	-150	-25	-150	-25	-25	-150	-25	-150	-25	-150	-25
A	100	100	-25	100	-25	100	100	-25	100	-25	100	-25	100
A#	-50	-25	-150	-25	-150	-25	-25	-150	-25	-150	-25	-150	-25
B	100	100	-25	100	-25	100	100	-25	100	-25	100	-25	100

	@	C	C#	D	D#	E	F	F#	G	G#	A	A#	B
C	1	1	0	1	0	1	1	0	1	0	1	0	1
C#	0	0	0	0	0	0	0	0	0	0	0	0	0
D	1	1	0	1	0	1	1	0	1	0	1	0	1
D#	0	0	0	0	0	0	0	0	0	0	0	0	0
E	1	1	0	1	0	1	1	0	1	0	1	0	1
F	1	1	0	1	0	1	1	0	1	0	1	0	1
F#	0	0	0	0	0	0	0	0	0	0	0	0	0
G	1	1	0	1	0	1	1	0	1	0	1	0	1
G#	0	0	0	0	0	0	0	0	0	0	0	0	0
A	1	1	0	1	0	1	1	0	1	0	1	0	1
A#	0	0	0	0	0	0	0	0	0	0	0	0	0
B	1	1	0	1	0	1	1	0	1	0	1	0	1

Fig. 1. Initialized "CMaj7 chord score table." The figure on the right is normalized.

3.2 Sequence Score

The sequence score tables determine points based on the melody patterns, and how many notes are in a bar, regardless of the harmony. In many parts of a jazz melody, a note or two move along the scale. Sometimes, the alignment of notes creates a parabola and jumps above the high pitches if required. Six patterns were picked out of several and used to teach the program to evaluate the points of a generated melody. Table 1 describes the pattern information in a sequence score table. "The number of notes and rests" refers to the number of notes in a bar including the rest notes. Because this study set the minimum note unit as an eighth note triplet, there can be up to 12 notes in a bar. "Between high and low" is the pitch difference between the highest and lowest notes in a bar. As the pitch difference is not much higher than 1 octave (12 pitches), the difference was regarded as 13 pitches. "The number of inflections" refers to the number of inflections in a bar. "Pitch difference between two notes" is the pitch difference between consecutive notes. Although it might be similar to the difference between consecutive notes in a chord score table, the chord score table is based on the syllable name, and therefore does not take pitch difference into consideration. Sometimes, there can be an octave (12 pitches) difference between two consecutive notes but a larger difference is uncommon; accordingly, we considered all differences greater than 13 pitches to be 13. "The number of lengths" is the duration information of notes or rests to be learned. The minimum note unit is an eighth note triplet and the note or rest duration can be established up to 12 (whole note); therefore, this value can range between 1 and 12. "The offset of start time" is the value at which notes or rests start. It represents 6-unit beats divided into two equal halves of one bar and can be used to learn the rhythm of the music.

Table 1. Elements of a sequence score table.

Row name of sequence score table	Range
The number of notes and rests	1–12
Between high and low	0–13
The number of inflections	0–10
Pitch difference between two notes	0–13
The number of length	1–12
The offset of start time	0–63

3.3   Learning and Evaluation

Given a song to be learned and a user rating, the developed program analyzes the pitch and pattern of the music and assigns a rating score to the corresponding attribute in the table. Figure 2 shows an example of this process.

These values are normalized for evaluation. The developed program searches for the attribute values of the test music based on the experience learned. If the number of values of each attribute is more than one, the average of values is the score of the attribute. An untrained attribute, "Jazz rhythm score," increases the score by 0.125 whenever it is at the beginning of the first and third triplet notes that are considered to fit well with the jazz rhythm. The sum of scores of all attributes is the fitness score. Figure 3 shows an example of evaluating a given music.

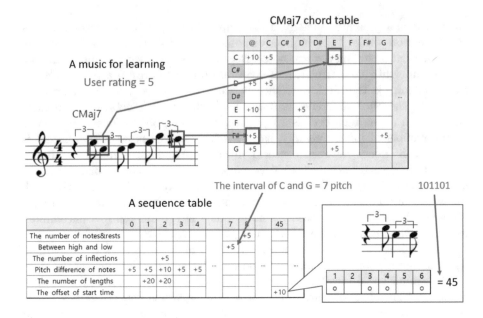

Fig. 2. Learning process of score tables.

4 Genetic Algorithm

4.1 Flow of the Algorithm

The flow of creating a piece of music using a GA in this study is illustrated in Fig. 4. Many popular melodies have their corresponding chord patterns. Chord patterns are already structured, and playing them well alone can produce a great piece of music. For a machine to produce all music elements automatically, it needs to create chord patterns. However, this study did not attempt this and focused only on the creation of melodies based on well-known chord patterns. A melody is a group of consecutive notes, and it is appropriate to express these notes as a chromosome of genes. The fitness function that determines the criteria for a good chromosome and geometric crossover that preserves a melody's good traits are key functions used in this study.

4.2 Representation

Many composition studies on GAs use the alignment of notes as a chromosome. It is not easy to convert all the notes in music into a single chromosome. There are numerous chords and notes in a piece of music, and it is difficult to determine which among them are good or bad considering only certain factors. As there is no clear equation that shows which alignment of notes is good, it is difficult to find a good schema. Accordingly, we set a space where a chromosome is expressed not as a score but as a bar; thus, a song consisting of 16 bars has 16 population

CMaj7 chord table(learned)

	@	C	C#	D	D#	E	F	F#	G
C	1	1	0	1	0	1	0.88	0	1
C#	0	0	0	0	0	0	0	0	0
D	0.2	0.51	0	0.51	0	0.57	0.77	0	0.38
D#	0	0	0	0	0	0	0	0	0
E	0.36	0.65	0	0.64	0	0.43	0.92	0	0.55
F	0.15	0.46	0	0.51	0	0.45	0.77	0	0.4
F#	0	0	0	0	0	0	0	0	0
G	0.45	0.78	0	0.51	0	0.51	1	0	0.4

A music for test

CMaj7

A sequence table (learned)

	0	1	2	3	4		7	8		45
The number of notes&rests		0.01	0.03	0.01	0.06		0.4	1		
Between high and low	0.06	0.02	0.06	0.3	0.16		0.83	0.51		
The number of inflections	0.39	1	0.98	0.55	0.15		0.01	0.01		
Pitch difference of notes	0.08	0.95	1	0.86	0.45		0.07	0.04		
The number of lengths		1	0.44	0.03	0.02		0.01	0.01		
The offset of start time	0.01	0.01	0.01	0.01	0.01		0.01	0.01		1

Fitness

Category	Fitness	Details
One note table	0.48	(1+1+0.2+0.36+0.36+0+0.45) / 7
Two notes table	0.61	(1+1+0.51+0.64+0+0.51) / 6
The number of notes&rests	1	one value
Between high and low	0.83	one value
The number of inflections	0.98	one value
Pitch difference of notes	0.72	(0.08+0.95+1+1+0.86+0.45) / 6
The number of lengths	0.72	(1+1+1+1+0.44+0.44+0.44+0.44) / 8
The offset of start time	1	(1+1) / 2
Jazz rhythm score	0.88	0.125 * 7
Total	7.22	

Fig. 3. Process of evaluating music.

groups. Notes contain both pitch and duration and each chromosome's length changes depending on the constituting notes. To represent this as a fixed length chromosome that can be easily handled by GAs, it is necessary to divide all of them into minimum unit beats. Then, the "hold flags" can be set to indicate that the notes are continuous. In this case, there are a variety of crossover techniques available. Figure 5 shows both cases where one bar is expressed as fixed length and variable length.

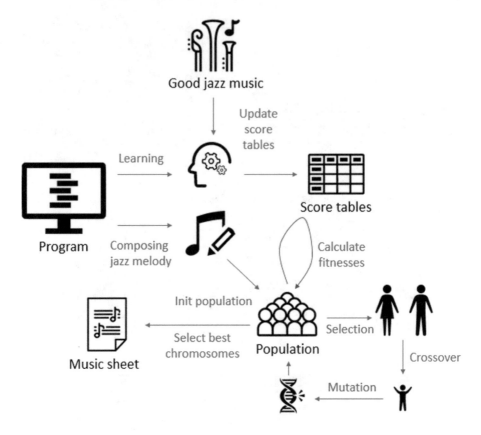

Fig. 4. Flow of automatic melody composition.

When a fixed length is used, there is an advantage that the attribute is unified into one and it is easy to perform GA operations. However, when crossover is performed, it is difficult to completely preserve the pitch and duration of good notes. When we use n-point crossover, if there is a point before the "hold flag," the duration of the note is reduced. This problem also occurs when uniform crossover [16] is applied. When a variable length is used, it has the advantage of preserving the pitch and duration of the parent note; however, designing GA operations is not easy. Fortunately, as shown in Sect. 3, the fitness function can evaluate chromosomes with variable length, and there is also a breakthrough technique to deal with variable-length chromosomes called a "geometric crossover." Hence, we chose to represent chromosomes with a variable length.

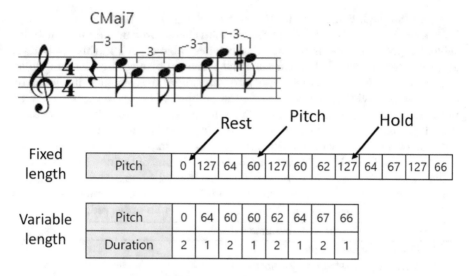

Fig. 5. Representation of a chromosome.

4.3   Crossover

The chromosome does not have a fixed-length pitch but a variable-length pitch to which a duration is assigned. Thus, geometric crossover, which can respond to variable lengths and preserve good traits of the parents well, was used. Geometric crossover is an operator that allows representation-independent recombination, making it possible to combine parents of different lengths. In addition, as it chooses genes based on the edit distance, it can consider a sequence, even though it selects the genes independently. Moraglio and Poli [17, 18] demonstrated that geometric crossover could be used well in biological sequences, and Yoon et al. [19] explained homologous crossover for variable-length sequences by using the mathematical unification of geometric crossovers. A melody is a sequence of notes and there is a distance between the notes in a musical space. This distance is the actual pitch difference; therefore, suitable musical transformation is required. The quality of music differs depending on how notes in music are recombined. Therefore, referencing the two studies mentioned above [17–19], crossover was designed to be applied to the notes in music.

Genes constituting a chromosome have two elements (pitch and duration). This results in a sub-match, which shows that notes can be partially identical; therefore, it must also be determined which element is more important. When good music flows well but there is lack of harmony, even those who do not know much about music find it odd. Because inappropriate pitch has a major effect on music, it should be removed. Duration is also an important factor; however, when the pitch sequence is good, changing each note's duration slightly does not have a major impact. Depending on which style is used to sing a song, singers change each note's duration intentionally. However, a melody's pitch is not changed unless it is a special case. Accordingly, it is highly important to preserve the pitch while the duration can be changed a little from time to time. To preserve a string's good elements, geometric crossover uses an edit distance table. To preserve good notes, an edit distance table is used in this study also, and the most optimized child chromosome is created by aligning a suitably filled table. Because a note has separate pitch and duration elements, it should be calculated by dividing the notes with the number of cases. Matching between notes with two elements can result in four cases (exact match, pitch match, duration match, and no match). Of course, in case of an exact match, the note must be preserved. Because of the difference in the importance of the pitch and duration explained above, in case of a pitch match, one of the notes is preserved. When notes only match in duration, they are treated as no match. This can be analyzed when aligning the edit distance table. Algorithm 1 shows the process of filling an edit distance table to select the pitch to be preserved.

Algorithm 1. Filling edit distance table

$//$ T: edit distance table
$//$ $Notes^1$, $Notes^2$: array of notes
$T_{i,0} \leftarrow i, T_{0,j} \leftarrow j$
for $i \leftarrow 1$ **to** $Notes^1.Length - 1$ **do**
    for $j \leftarrow 1$ **to** $Notes^2.Length - 1$ **do**
        if $Notes^1_{i-1} = Notes^2_{j-1}$ **then**
            $T_{i,j} \leftarrow T_{i-1,j-1};$
        else if $Notes^1_{i-1}.pitch = Notes^2_{j-1}.pitch$ **then**
            $T_{i,j} \leftarrow \min(T_{i-1,j-1}, T_{i-1,j}, T_{i,j-1}) + 1;$
        else
            $T_{i,j} \leftarrow \min(T_{i-1,j-1}, T_{i-1,j}, T_{i,j-1}) + 2;$

Figure 6 shows creation of an edit distance table through two parents notes on a score.

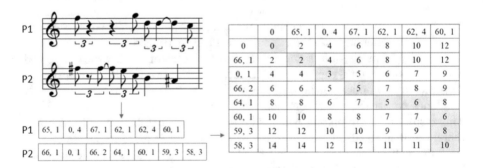

Fig. 6. Edit distance table for notes of parents.

A sub-match, which was not considered when completing an edit distance table, is analyzed now. The process of aligning a completed edit distance table is the same as the previous one, but has a slightly different meaning. As in a row-to-column table, a horizontal move is delete and a vertical move is insert. A diagonal move, however, does not replace but deletes and inserts a note when the number changes. One of the two notes must be selected when replaced. Instead, delete and insert give an option of removing both when the pitch changes. When the number does not change in a diagonal move, one of the two notes is selected randomly. When two notes (genes) are an exact match, one of the two notes is selected randomly and thus preserved. Accordingly, the analyzed results can be described as shown in Fig. 7. The transcript above is adjusted slightly from an expression used in the study of Moraglio and Poli [18]. (M) is a place holder and must be selected. Even though it is not described above, (R) is replace, which is indicated when there is a pitch match and one of the two notes is randomly selected. A gene is preserved in these two cases; selection is not made when insert (I) or delete (D) is performed. This can be illustrated as shown in Fig. 8.

P_1	-	65, 1	0, 4	-	67, 1	62, 1	62, 4	60, 1	-	-
P_2	66, 1	-	0, 1	66, 2	-	62, 1	-	60, 1	59, 3	58, 3
Edit	I	D	R	D	D	M	D	M	I	I

Fig. 7. Edit transcript associated with an alignment.

| Note | 66, 1 | 65, 1 | 0, 4 | 0, 1 | 66, 2 | 67, 1 | 62, 1 | 62, 4 | 60, 1 | 59, 3 | 58, 3 |
| Selected | F | T | F | T | F | T | T | T | T | T | F |

Fig. 8. Selected note table after place holding and random selection (T: selected, F: unselected).

The sum of the durations of the selected notes is 3, which less than the required duration of 12. Accordingly, one of the notes with F is selected randomly and turned into T until the sum of the durations of the selected notes become 12. Sometimes, the total duration exceeds 12 when notes continue to be selected; in this case, the exceeding duration value can be removed from the notes selected last. When this process is completed, crossover operation successfully generates a new chromosome that contains the notes selected as T (true).

4.4 Mutation

Mutation was performed in three parts. The first part was to change the pitch; according to the probability designated by the user, a random note's pitch is added or removed within the range of pitch designated by the user. Additionally, sharps were all removed because sharp (♯) notes do not belong to the C major scale. The second part was to remove random notes according to a designated probability. With the current crossover method, the number of notes increase slightly over generations; thus, if they are not removed, only eighth note triplets would remain after many generations. The third part made note changes between bars smooth. Currently, the pitch distance between genes within a chromosome is compared; unfortunately, there is no note comparison between chromosomes (bar) by the GA. Accordingly, a large pitch change can occur as a bar changes, and this function would bring a note closer to the last note of the previous bar or first note in the next bar. Therefore, it makes the bars transition more smoothly. Removing and smoothing notes is in fact similar to repairing even though performed using probability.

5 Experiment

5.1 Model Training

For valid fitness calculations, table values discussed in Sect. 3 should be learned first. For this, we need a lot of data related to the prepared jazz melodies with the same chord patterns. Fortunately, the "Band-in-a-Box" program discussed in Sect. 2 can create a solo melody for a given chord. However, this is based on an expert system and we observed that similar melodies were frequently produced. There is also a problem that swing rhythm is too well implemented and cannot be accepted in the learning system that requires accurate eighth note triplet time. Therefore, we changed the generated melody to shuffle rhythm and make small changes to the pitch. In this manner, 52 training data including 8 bars and melodies were prepared and the table was trained. Training data is available on the site[3].

5.2 Settings

The program was implemented using WinForms and Telerik[4]. WinForms is a graphical user interface (GUI) class library included as a part of Microsoft .NET framework [20]. An accompaniment was imported from the musical instrument digital interface file so that it could be played in the program. We let the program play a melody at the designated tempo and accompaniment. The hyperparameters of the GA can be modified in the program. The hyperparameters and experiment settings are described in Table 2.

---

[3] goo.gl/Ra8GeA.
[4] https://www.telerik.com.

Table 2. Experiment setting.

Category	Sub-category	Value
Initial random note	Pitch lower	3-octave A
	Pitch upper	6-octave C
	Duration lower	1
	Duration upper	12
GA	Population	100
	Roulette wheel k	4
	Mutation	0.1
	Elitism	1

5.3 Evaluation

The most important factor to be considered when assessing a composer is the quality of songs composed. We created a model that allows our program to judge this quality and calculate the score of the music; however, human evaluation is also very important in determining the quality. Therefore, the evaluation criteria have two phases: fitness value and human evaluation. Fitness value can be calculated by the program. However, because human evaluation standards are different, we should perform extensive human evaluation of our results. We obtained a rating on a melody that the machine composed through a Google survey on the Internet. The melodies to be evaluated by humans were 1st, 30th, 100th, 200th, 500th, and $1,000^{th}$ generations of the composed music. Figure 9 is an example of a melody after the 100th generation. This melody is available at the site[5].

5.3.1 Evaluation by Fitness Value

For monitoring the search process of our GA, the best and average fitness values were displayed in each bar. The fitness values evaluated in this study are the sum of the fitness values of all bars. Because one bar can have a maximum of 9 fitness values, the maximum fitness value of 16 bars becomes 144. We created 10 songs for 1,000 generations and recorded the best fitness value and the average value in the process. Figure 10 shows the best and average values of average fitness for 10 music evolutions. The standard deviation of the results over 10 runs was measured to not exceed 1.3 for all generations. Therefore, it is generally considered to be a reliable convergence. Moreover, because the fitness rises rapidly in the early stages, we can see that our geometric crossover operator on variable-length chromosomes successfully generates offspring.

---

[5] https://youtu.be/qElehTmTsoU.

Fig. 9. Music score at the 100th generation.

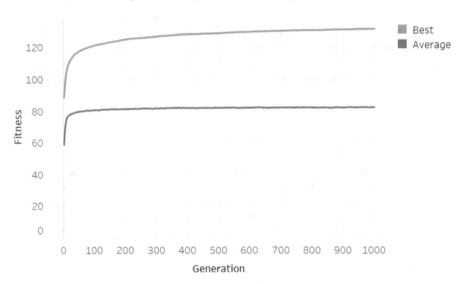

Fig. 10. Average of best fitness and average fitness while 10 music evolve.

5.3.2   Evaluation by Humans

It made sense to conduct a survey on the Internet because there is no locational constraint. However, we did not know who the respondents were; therefore, we evaluated the respondent's proficiency in music before evaluation. The results are shown in Fig. 11.

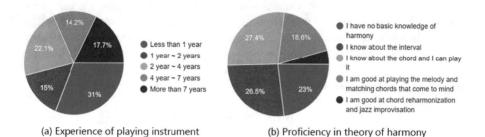

(a) Experience of playing instrument   (b) Proficiency in theory of harmony

Fig. 11. Proficiency in the music of respondents.

In the evaluation of music, respondents could assign a score of 1 (worst) to 5 (best) for each question. The content of the questions and average scores received from 113 people are shown in Fig. 12.

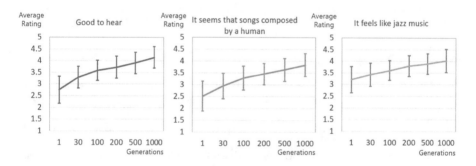

Fig. 12. Evaluation of music by humans.

Details of the response data can be found on the site[6]. It can be seen that music produced in the later generations is rated relatively high. It is interesting that some people who are skilled in music thought that music at early generations was similar to jazz. They commented that "over generations, the music sounded more conventional but less jazzy." As a result, we can confirm that the music produced by our GA is impressive to people as the generation progresses

6   Conclusions

We examined how to generate a jazz melody that fit the supplied chords using geometric crossover and learning score tables. This study demonstrated that

---

[6] goo.gl/QRbcxN.

parent notes' important sub-patterns can be learned, or passed down successfully. The model also learned the note pattern of training data and judged the fitness of the given music well. In our previous study, there was a critical problem of the production of similar melodies each time. This was due to attempts to prematurely find a better optimal point such as excessive mutation and weight concept in the edit distance table. To avoid this problem, we made the mutation operator weaker than that in the previous study, and simplified the edit distance table so that the music could fall into the appropriate local optima. This slowed down the rate of increase in fitness but increased the diversity of the melody.

Our method was successful in creating a jazz solo melody from fixed chords, but it does not produce a melody for untrained chords. In the future, we will generate more learning data and melodies from various chords. Moreover, the chord progression is not fixed but automatically created; we will develop a program for real automatic composing. We hope that this program will help composers come up with appropriate phrases when creating jazz music.

Acknowledgement. We would like to thank Prof. Francisco Fernández de Vega for providing us with much advice and help in writing this paper. This research was supported by a grant [KCG-01-2017-05] through the Disaster and Safety Management Institute funded by Korea Coast Guard of Korean government, and it was also supported by Basic Science Research Program through the National Research Foundation of Korea (NRF) funded by the Ministry of Education (No. 2015R1D1A1A01060105).

References

1. Montag, C., Reuter, M., Axmacher, N.: How one's favorite song activates the reward circuitry of the brain: personality matters! Behav. Brain Res. **225**(2), 511–514 (2011)
2. van Eijck, K.: Social differentiation in musical taste patterns. Soc. Forces **79**(3), 1163–1185 (2001)
3. Papadopoulos, G., Wiggins, G.: AI methods for algorithmic composition: a survey, a critical view and future prospects. In: AISB Symposium on Musical Creativity, vol. 124, pp. 110–117 (1999)
4. Biles, J.A.: Genjam: a genetic algorithm for generating jazz solos. In: Proceedings of the International Computer Music Association, vol. 94, pp. 131–137 (1994)
5. Matić, D.: A genetic algorithm for composing music. Yugoslav J. Oper. Res. **20**(1), 157–177 (2010)
6. Horner, A., Goldberg, D.E.: Genetic algorithms and computer-assisted music composition. Urbana **51**(61801), 437–441 (1991)
7. Quick, D.: Kulitta: A Framework for Automated Music Composition. Yale University (2014)
8. Keller, R.M., Morrison, D.R.: A grammatical approach to automatic improvisation. In: Proceedings of the Fourth Sound and Music Conference (2007)
9. Bickerman, G., Bosley, S., Swire, P., Keller, R.M.: Learning to create jazz melodies using deep belief nets. In: Proceedings of the International Conference on Computational Creativity, pp. 228–237 (2010)
10. Johnson, D.D., Keller, R.M., Weintraut, N.: Learning to create jazz melodies using a product of experts. In: Proceedings of the International Conference on Computational Creativity (2017)

11. Fernández de Vega, F.: Revisiting the 4-part harmonization problem with GAs: a critical review and proposals for improving. In: Proceedings of the IEEE Congress on Evolutionary Computation, pp. 1271–1278 (2017)
12. Nam, Y.-W., Kim, Y.-H.: A geometric evolutionary search for melody composition. In: Proceedings of the Genetic and Evolutionary Computation Conference Companion, pp. 53–54 (2018)
13. Nam, Y.-W., Kim, Y.-H.: Melody composition using geometric crossover for variable-length encoding. In: Proceedings of the Genetic and Evolutionary Computation Conference Companion, pp. 37–38 (2017)
14. Johnson, D.D.: Generating polyphonic music using tied parallel networks. In: Correia, J., Ciesielski, V., Liapis, A. (eds.) EvoMUSART 2017. LNCS, vol. 10198, pp. 128–143. Springer, Cham (2017). https://doi.org/10.1007/978-3-319-55750-2_9
15. Hadjeres, G., Pachet, F.: Deepbach: a steerable model for bach chorales generation. In: Proceedings of the 34th International Conference on Machine Learning, no. 70, pp. 1362–1371 (2017)
16. Syswerda, G.: Uniform crossover in genetic algorithms. In: Proceedings of the Third International Conference on Genetic Algorithms, pp. 2–9 (1989)
17. Moraglio, A., Poli, R.: Geometric landscape of homologous crossover for syntactic trees. Evol. Comput. 1, 427–434 (2005)
18. Moraglio, A., Poli, R., Seehuus, R.: Geometric crossover for biological sequences. In: Collet, P., Tomassini, M., Ebner, M., Gustafson, S., Ekárt, A. (eds.) EuroGP 2006. LNCS, vol. 3905, pp. 121–132. Springer, Heidelberg (2006). https://doi.org/10.1007/11729976_11
19. Yoon, Y., Kim, Y.-H., Moraglio, A., Moon, B.-R.: A mathematical unification of geometric crossovers defined on phenotype space. arXiv:0907.3200 (2009)
20. Sells, C., Weinhardt, M.: Windows Forms 2.0 Programming (Microsoft Net Development Series). Addison-Wesley Professional, Boston (2006)

Exploring Transfer Functions in Evolved CTRNNs for Music Generation

Steffan Ianigro$^{(\boxtimes)}$ and Oliver Bown

UNSW Art and Design, Paddington, NSW 2021, Australia
steffanianigro@gmail.com, o.bown@unsw.edu.au

Abstract. This paper expands on prior research into the generation of audio through the evolution of Continuous Time Recurrent Neural Networks (CTRNNs). CTRNNs are a type of recurrent neural network that can be used to model dynamical systems and can exhibit many different characteristics that can be used for music creation such as the generation of non-linear audio signals which unfold with a level of generative agency or unpredictability. Furthermore, their compact structure makes them ideal for use as an evolvable genotype for musical search as a finite set of CTRNN parameters can be manipulated to discover a vast audio search space. In prior research, we have successfully evolved CTRNNs to generate timbral and melodic content that can be used for electronic music composition. However, although the initial adopted CTRNN algorithm produced oscillations similar to some conventional synthesis algorithms and timbres reminiscent of acoustic instruments, it was hard to find configurations that produced the timbral and temporal richness we expected. Within this paper, we look into modifying the currently used *tanh* transfer function by modulating it with a *sine* function to further enhance the idiosyncratic characteristics of CTRNNs. We explore to what degree they can aid musicians in the search for unique sounds and performative dynamics in which some creative control is given to a CTRNN agent. We aim to measure the difference between the two transfer functions by discovering two populations of CTRNNs using a novelty search evolutionary algorithm, each utilising a different transfer function. The effect that each transfer function has on the respective novelty of each CTRNN population is compared using quantitative analysis as well as through a compositional study.

Keywords: Continuous Time Recurrent Neural Network ·
Novelty search · Audio synthesis · Generative music

1 Introduction

In prior work [1–3], we have investigated the discovery of musical Continuous Time Recurrent Neural Networks (CTRNNs) using Evolutionary Algorithms (EAs). These papers revolve around harnessing the unique characteristics of CTRNNs for electronic music creation using EAs as a means to discover their

© Springer Nature Switzerland AG 2019
A. Ekárt et al. (Eds.): EvoMUSART 2019, LNCS 11453, pp. 234–248, 2019.
https://doi.org/10.1007/978-3-030-16667-0_16

breadth of behaviours. The key interest in the use of CTRNNs comes from their potential for creative search. They offer a huge space of audio possibilities which is explorable by manipulating a finite set of parameters, making them ideal as a genotype structure for an EA. Furthermore, the non-linear capabilities of CTRNNs can create a large range of different temporal behaviours which can be used for electronic music creation. For example, a CTRNN could be used to synthesise audio or be used as a generative or metacreative [4] agent that has creative influence over the rhythmic structure and melodic movements of a work. CTRNNs can also manipulate their input signals in real time, making them appealing candidates for digital signal processing (DSP) in interactive music-making contexts such as live laptop performance.

During our initial studies, we encountered varying degrees of success when discovering CTRNNs for electronic music creation. Although CTRNN configurations were discovered that exhibited a range of timbres, we were not satisfied with the algorithm design due to the overall lack of long-term non-linear behaviours that would be of most interest to musicians. For example, CTRNNs were discovered that exhibit sonic qualities of clarinets and saw-tooth waveforms, but could not produce musical variance over a large scale such as gradually changing rhythmic and melodic motifs nor exhibit significant levels of surprise when interacting with them in live performative contexts.

The aim of this paper is to explore a different CTRNN activation function to encourage extended melodic and dynamic variance in a CTRNN's output. As a result, the musical application of CTRNNs could shift from usage as a synthesis algorithm to a co-creative performance situation in which a musician can interact with a CTRNN to produce a range of musical structures that have a richness imparted by its idiosyncratic characteristics. For example, a CTRNN could be used to control meta-structures of a work or generate subtle rhythmic components, influencing a final creative artefact beyond the direct influence of a human musician. The compositional study outlined in this paper is an example of this, as a CTRNN is used to manipulate a guitar signal to enhance its dynamic and timbral qualities in indirect ways. This is interesting from a performative context as, differing from many existing guitar effects which often have a very linear relationship between input and output, a CTRNN can often fluctuate to different sonic states in ways often abstract from the moments of its input. Each output of a CTRNN can also exhibit different, yet correlated behaviours, providing possibilities for rich one-to-many, many-to-many or many-to-one mappings of audio signals. So far, in our prior work, we have not been able to find CTRNNs that exhibit these more complex characteristics. However, implementation of a new transfer function could help enhance the behaviours that we have discovered in prior work.

The second author has explored the possibility of modifying the activation function of a CTRNN to encourage more complex behaviour. This modification consists of modulating a standard *tanh* transfer by a *sine* function, shifting a CTRNN's functionality from favouring more stable behaviours to encouraging more expressive performance dynamics and varied network interactions. Furthermore, the new transfer function, which we refer to as the *tanhsine* function, will also have two additional genetic parameters for each neuron that can be evolved to explore how various degrees of this *sine* modulation impacts the musical behaviour of a CTRNN.

Comparison of these two transfer functions is conducted through an experiment and compositional study based on two CTRNN populations discovered using the NS algorithm described in [3], one consisting of CTRNNs using the *tanh* transfer function and the other consisting of CTRNNs using the *tanhsine* transfer function. Specifically, graphical representations of the CTRNN populations are discussed, providing a quantitative comparison of the discovery rates for each transfer function. Additionally, the compositional study provides a qualitative analysis to help understand how the two different transfer functions impact the practical application of CTRNNs to electronic music composition. The audio visual study can be found at the link: https://vimeo.com/300107683/0ac8c3ab04.

This compositional process consists of selecting the most musically interesting individual from both novel CTRNN populations using the online appendix for this project (http://evomusart.plecto.xyz) as well a CTRNN Max For Live device that can be found in the online repository for this project (https://github.com/steffanianigro/plecto). The selected CTRNN is then used as a type of guitar effect to enhance a guitar signal, providing insight into the range of musical characteristics a CTRNN can impart. A visual component is also added to the work, further emphasising the CTRNN dynamics at play. These analytical approaches will highlight the differences between the two transfer function implementations when applied to creative domains and provide an insight into the nuances of creative interactions with these CTRNNs when using them to produce electronic music.

We found that the *tanhsine* transfer function produced more temporal and timbral variety than the *tanh* transfer function. The temporal variation produced by the *tanh* transfer function was often repetitive and fairly stable. In contrast, the *tanhsine* transfer function produced richer dynamics such as gradual, non-repeating fluctuations or drastic audio state shifts in a CTRNN's output. Furthermore, the *tanhsine* transfer function was more usable in musical contexts, providing a malleable generator of musical structure that can potentially be applied to many different creative practices.

2   Background

2.1   CTRNN

Artificial Neural Networks (ANNs) have been used in many different systems for the creation and exploration of audio waveforms. *Wavenet* [5] is a notable example, in which deep neural nets are used to create raw audio, such as the generation of speech from text and music audio-modelling. Kiefer's use of Echo State Recurrent Neural Networks for instrument mapping [6] is another example, capitalising on non-linear behaviours to achieve interesting musical interaction. Mozer adopts a higher level approach in which ANNs become artificial composers that can extract stylistic regularities from existing pieces and produce novel musical compositions based on these learnt musical structures [7]. Lambert et al. provide an example of how Recurrent Neural Networks can be used as a beat tracking algorithm [8]. Additionally, Continuous Time Neuron Models are explored by Eldridge [9] for their properties as responsive mechanisms for live interactive performances. We adopt CTRNNs for similar properties as they are capable of producing a variety of temporal dynamics that could be used for audio synthesis, control signal generation and the creation of higher level musical structures.

The CTRNN is an interconnected network of computer-modelled neurons capable of a rich range of dynamical behaviours [10]. Bown and Lexer provide an example of these dynamics by using CTRNNs as musical agents to generate rhythmic gestures [11], work which we use as the basis for the CTRNN algorithm described in this paper. Through explorations of the CTRNN algorithm, we have found that it can also be applied to the audio signal generation domain, with each iteration of a CTRNN producing a sample in an audio waveform. Their non-linear continuous dynamics result in various waveforms that can be controlled by manipulating each node's parameters - gain, bias and time constant - and connection weights. Due to the complex relationship between these network parameters and their effects of the network output, CTRNNs are often trained by an EA [11].

Each neuron of a CTRNN is typically of a type called the *leaky integrator*. The internal state of each neuron is determined by the differential equation 1,

$$\tau_i(dy_i/dt) = -y_i + \sum W_{ij}\sigma(g_j(y_j - b_j)) + I_i \qquad (1)$$

where τ_i is the time constant, g_i is the gain and b_i is the bias for neuron i. I_i is any external input for neuron i and W_{ij} is the weight of the connection between neuron i and neuron j. σ is a *tanh* transfer function [11].

Further to the *tanh* transfer function, within this paper we also utilise a *tanhsine* transfer function determined by Eq. 2,

$$y = (1 - c)\tanh(a) + c\sin(fa) \qquad (2)$$

where c is a sine coefficient, a is the activation of the neuron and f is a frequency multiplier. The behaviour of the sine coefficient variable changes the strength of the *tanh* function's modulation by the *sine* function and the frequency multiplier variable controls the frequency of the *sine* function. Below is a graphical comparison of the *tanh* (Fig. 1) and *tanhsine* (Fig. 2) transfer functions.

2.2 Evolutionary Algorithm

Evolutionary Algorithms (EAs) have been used widely by musicians as a means to explore a creative search space for desirable artefacts. Examples of this work include: the interactive evolution of synthesisers/synthesis algorithms using Dahlstedt's *MutaSynth* [12]; the sonified dynamics of an ecosystem model in

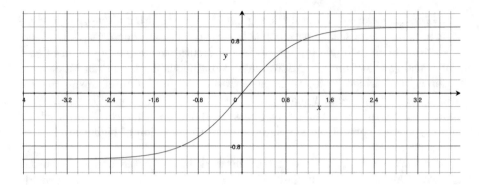

Fig. 1. *tanh* transfer function.

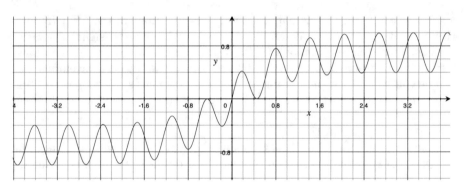

Fig. 2. *tanhsine* transfer function, where $c = 0.3$ and $f = 10$.

McCormack's *Eden* [13]; evolution of music by public choice in *Darwin Tunes* [14] by McCallum et al.; the creation of automated improvisers using timbral matching [15]; and Bown's evolution of metacreative black boxes to create improvising companions [16,17].

In prior works [1–3], we have investigated interactive, target-based and novelty search algorithms to explore the CTRNN musical search space, with results varying with each experiment. Interactive evolutionary approaches were often inhibited by fitness bottlenecks in which aesthetic evaluation of the population slowed down the search process [18]. Target-based approaches mitigated this issue as CTRNNs were evolved towards a computer-encoded target but did not allow the same control over aesthetic trajectory afforded by interactive evaluation. Novelty search (NS) provided the best of both worlds as the algorithm design does not require a pre-defined fitness objective and instead simply looks for novelty among individuals. This means the search process is not constrained by strict criteria that may not even provide the best path to an optional solution, facilitating open-ended discovery [19]. Furthermore, human evaluation is not required as part of the evolutionary process, speeding up the rate of search space discovery. Use of the NS algorithm also affords the discovery of a large population containing novel individuals, each with unique characteristics that can be harnessed using other search methods depending on a musician's creative use case. For example, we are in the process of creating an online system which allows users to find musically applicable CTRNNs from a population discovered using NS. In this case, the user is extrapolated from the tedious process of sifting through a large array of non-musical CTRNNs and instead, traverse through a refined population that can be represented or filtered in many different ways to suit the user's creative purpose.

In this paper, we use an *opt-aiNet* algorithm based on the design by de Castro and Timmis [20]. This algorithm has been further explored in works such as [21] and [22]. In [3], we adopted this algorithm for novelty search due to its capabilities regarding multi-modal optimisation. The algorithm allows individuals to be evolved in isolated sub-populations using a range of independent metrics and maintains pareto optimal candidates through each iteration of the evolutionary process. The algorithm also implements various population suppression and diversity mechanisms, meaning if the algorithm starts to stagnate, it will remove similar individuals and seed new members of the population to mitigate divergence on a local optimum.

3   Experiment

In this experiment, we aimed to discover as many novel CTRNN configurations (two input nodes, six hidden nodes and two output nodes) as possible over 300 iterations using both the *tanh* and *tanhsine* transfer functions. As the NS process can be computationally expensive due to the rendering and analysis of audio waveforms, this process took about 48 h for each transfer function. All NS parameters were exactly the same for each experiment including the novelty thresholds (a multi objective NS algorithm was used). We adopted two metrics to measure novelty in the *opt-aiNet* NS algorithm. Mel-Frequency Cepstral Coefficients (MFCCs) were used to measure the timbral difference between CTRNN configurations such as described in [3]. Additionally, a melodic variance metric was used to differentiate individuals that exhibited spectral change over longer periods of time. The variance metric used was calculated by analysing multiple windows of a piece of audio for pitch content by extracting the dominant frequency from an FFT spectrum. We only used the dominant frequency as in this experiment we were only concerned with broader melodic variance in a CTRNN's output. In future work, we will look to introduce other spectral measures to refine this calculation. Once the frequency content of each window was identified, a moving variance algorithm was used to measure the cumulative change of the dominant frequency values. This final value was compared within the existing population to attain how novel an individual's temporal melodic variance was. About 12 s of audio was rendered for each output of a CTRNN and was broken into three overlapping audio windows for analysis. When calculating the MFCC metric, the three FFT spectrum were averaged to reduce noise prior to the calculation (this may change in future work to allow for detection of timbral change over time). Large audio windows were used to identify broader melodic gestures when calculating the melodic variance between windows.

In this experiment, the CTRNN inputs used in the *opt-aiNet* NS algorithm were static values ranging from −1 to 1. We opted for static input values instead of more complex waveform structures when assessing the novelty of a CTRNN configuration as we wanted to ensure any timbral or longer term temporal behaviours were a result of a CTRNN itself and not filtered or amplified aspects of its input. For example, if two sine waves were used as a CTRNN's inputs, inter-action of these signal within the CTRNN could cause new timbres or rhythmic components that may have little to do its internal dynamics.

One of the challenges of evolving CTRNNs is that their outputs vary greatly when fed different input structures. For example, if a constant value of 1 is fed into a CTRNN, it may produce continuous oscillations. However, if this value is changed, the CTRNN may suddenly saturate and produce a fixed value of 1 or −1. To avoid CTRNNs that fall into troughs of non-musical behaviour, we opted for the use of multiple different sets of inputs when evaluating a single CTRNN. Specifically, four sets inputs (two input values each) were used to generate four sets of outputs (two output values each). The novelty metrics were calculated for each output of each input iteration, providing a total of sixteen novelty indicators per CTRNN (two metrics for each of the two CTRNN outputs over four input iterations). These indicators were all compared individually and were treated as multiple novelty objectives for the *opt-aiNet* NS algorithm. This approach was adopted as we did not want to smooth over any CTRNN idiosyncrasies by averaging metrics. For example, if a CTRNN produced a very novel output when fed a specific input but did not do anything interesting when fed other input types, averaging its metric values could indicate that the CTRNN is just as novel as another CTRNN that offers a small amount of novelty for each input type. Yet if these CTRNNs were to be used in the real world, they will both offer very different experiences as one requires a very specific circumstance in which to produce musical novelty. In contrast, when using multiple novelty objectives, the first CTRNN will only gain a high novelty rank for one of the sixteen novelty objectives and a low rank for the rest, reducing its novel standing within the population and representing its behaviour in the real world.

The graph in Fig. 3 shows the discovery rates for each transfer function. The results show a large difference between the two CTRNN implementations and as predicted, the use of the *tanhsine* transfer function greatly improved the discovery rate of CTRNNs that are novel in both the timbral and variance domains. The source code for the *opt-aiNet* NS algorithm can be found here: https://github.com/steffanianigro/plecto.

3.1   Discussion

We can see that after 300 iterations of the *opt-aiNet* NS algorithm, the number of resulting individuals differs greatly with each transfer function. 754 novel CTRNNs where discovered using the *tanhsine* transfer function compared to

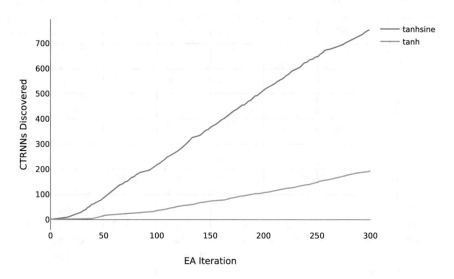

Fig. 3. Discovery rate comparison between *tanhsine* and *tanh* transfer functions. Note that discovery rates will always depend on the novelty metrics used. However, we assume that the *tanhsine* transfer function population is likely to be more varied as the transfer function used is more complex.

the 194 novel CTRNNs discovered using the *tanh* transfer function. This demonstrates the enhanced capabilities of the *tanhsine* transfer function for the generation of novel timbral and temporal characteristics. It is worth noting that when the algorithm is run with only the MFCC timbral metric to measure novelty, the difference between the two transfer functions becomes less noticeable as both the *tanhsine* and *tanh* transfer functions are capable of a large breadth of timbres. However, the main musical interest of CTRNNs lies in their ability to not only produce a range of timbres but exhibit long-term temporal changes, a behaviour which is more evident when using the *tanhsine* transfer function. In future work, we will further refine the novelty thresholds and metrics to broaden the distance between novel individuals as Fig. 3 indicates that the algorithm could potentially find many more novel individuals with little indication that the discovery rate is tapering off. Considering the current population size, if many more individuals are discovered, the population may exceed the limits of discoverability by a musician depending on the user interface provided.

We have also found that the large number of novelty objectives used within this experiment can cause optimisation issues which we will look to solve in further iterations of the algorithm. For example, the algorithm has to render approximately 96 s of audio for each CTRNN (12 s for both outputs of each of the four input iterations) which has large computational performance costs. In pursuit of a solution, we have observed that if a single CTRNN output exhibits novel behaviour, we can often assume that its other outputs will be similarly varied as each CTRNN node often has some behavioural correlation. Therefore, we could reduce the number of novelty objectives by only analysing a single output of a CTRNN when assessing novelty instead of the two currently used. CPU parallelisation was also applied to the NS algorithm to increase efficiency and will be refined in future work.

This experiment has shown the variation between CTRNNs using a quantitative approach. However, these charts do not represent the usability of these CTRNN configurations in musical contexts. For example, a CTRNN utilising the *tanhsine* transfer function may exhibit extensive timbral and temporal variance yet could be too inconsistent or temperamental to be useful to a musician. Therefore, further to charting the discovery rates of the *opt-aiNet* NS algorithm, we have also adopted a qualitative approach to the analysis of the two CTRNN populations, providing an understanding about how each transfer function affects the musicality of a CTRNN and its ability to integrate into existing creative workflows. This is achieved by conducting a compositional study which explores the musical affordances of the *tanh* and *tanhsine* CTRNN transfer functions.

We expect to find that the *tanhsine* transfer function will provide a larger range of dynamic variety over the *tanh* transfer function. Additionally, we expect that the *tanhsine* transfer function will encourage sudden state changes in a CTRNN's output compared to use of the *tanh* transfer function which will produce more consistent behaviour. Finally, we expect that the *tanhsine* transfer function will be more responsive to a CTRNN's input, meaning both static and complex structures can be fed into a CTRNN and result in obvious shifts in the CTRNN's output. In contrast, we predict the *tanh* transfer function will be more rigid in its response to different input types, either changing very slightly or not at all. As a guitar signal wasn't used as the input for the *opt-aiNet* NS algorithm CTRNNs, we expect that there will be a consistency around un-responsiveness to the subtle movements of a guitar input as the CTRNNs were trained on static inputs. Even though static inputs were necessary to ensure the CTRNNs themselves can produce novel behaviour, this may cause issues when using them in more dynamic contexts.

4 Compositional Study

The compositional study component of this paper further investigates the different behaviours exhibited by the *tanh* and *tanhsine* transfer functions and how they can be used to manipulate a guitar input signal in non-linear ways. This process consisted of the identification of one configuration from both populations (one population using the *tanh* transfer function and the other population using the *tanhsine* transfer function) of novel CTRNNs that best fit our goal of creating an audio-visual work using guitar and CTRNN. The selection criteria revolved around the identification of CTRNNs that can transform the timbral and structural content of a guitar input signal in real time whilst maintaining some aspects of the input audio. The appendix for this paper (http://evomusart. plecto.xyz) and a Max For Live device that integrates into Ableton Live [23] (https://github.com/steffanianigro/plecto) were used to preview the CTRNNs and find the one most suitable for this study. The final work consists of audio generated by the selected CTRNN combined with Lissajous visualisations of the two CTRNN outputs (output one plotted to the x-axis and output two plotted to the y-axis). Lissajous visualisations are used in this study as not only do they look appealing but they also aid in the understanding of the non-linear behaviours a CTRNN can produce.

From this study, we aimed to identify the transfer function that is most conducive to creative application. This was achieved by discussing both the exploratory process of finding the ideal individuals from each population of novel CTRNNs as well as the process of using the selected CTRNN to create a small audio-visual guitar work. An online appendix to this study which highlights the differences between individuals from both the *tanh* and *tanhsine* populations can be found here: http://evomusart.plecto.xyz.

A technical description of the setup used to create the small composition is as follows: the first author has implemented a Max/MSP [24] CTRNN object that allows CTRNN configurations to be dynamically loaded and played at audio rate. This object has been used to create a Max For Live (M4L) device that integrates into the popular Digital Audio Workstation Ableton Live [23]. This M4L device feeds into another M4L device which utilises Jitter [24] to create Lissajous visualisations of the left and right outputs of the CTRNN M4L device that correlate to the first and second outputs of the encapsulated CTRNN. A guitar with a clean tone was used as the M4L device input (the same guitar signal was fed into both CTRNN inputs). An online repository consisting of the CTRNN libraries used in the study can be found here: https://github.com/steffanianigro/plecto.

4.1 Search Process

We identified three key differences between the *tanh* and *tanhsine* transfer functions during our navigation of the novel CTRNN configurations discovered by the *opt-aiNet* NS algorithm: the range of dynamics the two transfer functions were capable of producing, the differing capabilities for shifting between states regardless of input change, and the responsiveness of a CTRNN to input movement.

Dynamic variance was an important feature that we were looking for during the creation of an audio-visual work, facilitating rhythmic events and long-term structural movement. We found that the CTRNNs using the *tanh* transfer function often produced waveforms with a limited range of dynamic attack and decay structures, resulting in many different static oscillations. In contrast the *tanhsine* transfer function produced much more dynamic variety, resulting in sounds with more accentuated dynamic forms and rhythmic and structural content, from subtle beating to sharper percussive forms. This was an interesting difference as once the *tanh* CTRNNs stabilised, they did not change unless their inputs were adjusted, making it difficult to introduce long-term dynamic movements in a work. In contrast the *tanhsine* CTRNNs would often produce turbulent dynamic movement such as constantly changing rapid rhythms or larger-scale state changes from soft textures to loud synthesiser-like tones, opening up the scope for creative expression. However, we also found that the more unstable nature of the *tanhsine* sometimes resulted in constant shifts between loud noise and silence that were not aesthetically pleasing.

In addition to the creation of dynamic variance, the ability for a CTRNN to smoothly move between textural spaces with little provocation was required. In this behaviour, the *tanhsine* transfer function outperformed the *tanh* transfer function, with CTRNNs often moving in and out of sonic states independent of changes to their inputs. For example, the *tanhsine* CTRNNs started producing continuous oscillations and then gradually introduced additional spectral content, often similar to that of electric guitar feedback. Sometimes this shift would be very extreme, quickly moving from sine like oscillations to noise, meaning the CTRNNs will be less usable if more subtle behaviour is required. However, the *opt-aiNet* NS algorithm provided various permutations of this behaviour, including more subtle, gradual shifts. In contrast, the *tanh* CTRNNs tended to remain in consistent states and did not shift much unless their inputs were changed, in which case the shifts were often very sudden and not very useful. This can be problematic from a musicality perspective as there are many readily available continuously oscillating algorithms that can be easier to control and manipulate than a CTRNN. The real interest of a CTRNN lies in its non-linear characteristics which were much more accessible when the *tanhsine* transfer function was used. This made the experience using the *tanhsine* transfer function much more enjoyable to create music with as there was a component of chance or surprise that often produced engaging musical results. The *tanh* CTRNNs also produced some level of surprise but in general were much more stable to work with as once they started producing a certain sound, they would continue to do so until agitated.

Consistent control of a CTRNN's output is another difference that we observed between transfer function implementations. This differentiation was particularly important as a user of a CTRNN audio processor should be able to find a configuration that allows varied levels of control depending on their creative application. For example, a musician may require a performative level of control over a CTRNN or they may be building an installation in which they may wish to capitalise on a CTRNN's autonomous capabilities. We found that the *tanh* CTRNNs would either be hard to control or act as a simple distortion or filter, with few configurations affording more complex non-linear behaviours that can be easily manipulated by their inputs to integrate with existing musical workflows. In contrast, the *tanhsine* CTRNNs seemed more responsive, with manipulation of CTRNN inputs resulting in correlated fluctuations on a more consistent basis. Furthermore, the *tanhsine* transfer function also produced a situation in which the input of a CTRNN had a meta-level of control over its output as opposed to the more direct nature of interaction observed with the *tanh* CTRNNs. For example, a small change to a CTRNN's input will flip its state completely or change its timbre in richer ways. As a result, the creative interaction was more interesting, expanding the scope of use beyond that of a simple audio effect. However, this meta level of control did introduce usability issues as sometimes the CTRNNs were too erratic to use effectively. Yet, it could be argued that if direct control is required over an audio generation tool, there may be better options than a CTRNN which shows more promise when used for its non-linear, generative features.

4.2 Compositional Outcomes

The compositional study has confirmed our initial insights into the impacts of the *tanhsine* and *tanh* transfer functions on a CTRNN's musicality regarding both the breadth of behaviours they can produce and how usable they are in compositional workflows. A link to the audio-visual work created can be found below and uses a CTRNN with the *tanhsine* transfer function. This CTRNN was selected as it manipulated its guitar input by introducing dynamic and melodic content, it was responsive to subtle changes in its input and it preserved various nuances of the input guitar's timbre. Specifically, various harmonic content was introduced into the spectrum when certain notes on the guitar were played, often contributing richer timbral content and obscure harmonic direction. Subtle rhythms or beating was also exaggerated by the CTRNN, creating some interesting rhythmic overlays and subtle feedback-like tones. Of most interest were the non-linear state changes from clean guitar timbres to more noise-like qualities that fluctuated depending on the intensity and frequency content of the guitar input, often exaggerating small dynamic and pitch movements.

https://vimeo.com/300107683/0ac8c3ab04

When navigating the various novel individuals in the *tanh* transfer function population, it was very difficult to find individuals that had similar behaviours. Many individuals either didn't respond to the guitar input or removed any presence of the guitar's qualities. Navigation of the *tanhsine* transfer function population was also a bit restricting at times for opposite reasons. Some of the configurations would be overly responsive and just produce noise whenever a change in a CTRNN's input occurred. Similarly to the *tanh* individuals, some configurations refused to respond to the input guitar movements which is not overly surprising due to the inputs used during the discovery of the CTRNNs. This was confirmed when running the same CTRNN configurations using flat signals as they responded better to input fluctuations. However, overall there were many more potential CTRNN candidates for the compositional study in the *tanhsine* population, indicating that in future work, the CTRNN algorithm should be modified to adopt these new changes when applied to musical domains.

5 Conclusion

Within this paper we have expanded on our prior research into the use of evolved CTRNNs as generator and manipulator of audio. Specifically, we argue that modification of the standard *tanh* transfer function used in prior research to incorporate *sine* modulation enhances the musical possibilities of the CTRNN, provoking richer behaviours. We have demonstrated this using an *opt-aiNet* NS algorithm to evolve two populations of CTRNNs over 300 algorithm iterations, one using the *tanh* transfer function and one using the modified *tanhsine* transfer function. The results of this experiment support our claim, with the discovery rate of the NS algorithm increasing when the *tanhsine* transfer function was adopted. Further to this quantitative evaluation, a qualitative analysis was conducted on each of the populations discovered. This analysis consisted of a small compositional study in which the two populations of CTRNNs were navigated by the authors to identify a CTRNN that can be used as a type of non-linear guitar effect. An audio-visual work was created using this CTRNN consisting of the direct output of the CTRNN combined with Lissajous visualisations. This creative process helped us understand the differences between adoption of the two transfer functions in regard to musicality and how they behave when integrated into a compositional workflow. Overall, we feel that the *tanhsine* transfer function is a more plausible choice when applying CTRNNs to musical contexts as it produced more dynamic variation, increased the scope for non-linear behaviours and encouraged engaging and surprising relationships between a CTRNN's input and output.

In future work, we aim to modify the *opt-aiNet* NS algorithm metrics to include more extensive variance detection than just melodic movement. This will help differentiate more subtle rhythms, dynamic fluctuations and larger structural changes within a piece of audio. As we further use the algorithm in creative contexts, metric refinements will also aid in the identification of CTRNN configurations that can serve specific musical tasks instead of broader behaviours. This will help mitigate exposure to CTRNN configurations that are too rigid or chaotic for musical purposes. We also aim to implement different input structures in the NS algorithm design to encourage CTRNN behaviours that are more responsive to a range of input types. This will improve some of the usability issues resulting from the rigid, input-independent behaviours observed as a result of the static input structures used in the *opt-aiNet* NS algorithm.

The efficiency of the algorithm has also caused optimisation issues as each iteration takes a long time due to the required audio analysis components. Therefore, we also hope to optimise this process by reducing the number of novelty objectives, thus the amount of audio processing required. Further refinements to the parallelisation of the NS algorithm processes using CPU or GPU optimisation may also be advantageous.

The user interface component of this research is also yet to be discussed in detail such as how the novel populations discovered by the *opt-aiNet* NS algorithm can be effectively navigated by musicians. In this paper we used a basic approach guided by some of the metrics extracted in the NS process. Future enhancements and additions to these metrics will enable more in-depth navigation and filtering of the novel populations to find key areas of musical interest. We aim to implement this interface ready for public use in the coming months and release the Max/MSP object and Max For Live device so that the broader community can engage in this project. An online repository for this project can be found at https://github.com/steffanianigro/plecto.

References

1. Ianigro, S., Bown, O.: Plecto: a low-level interactive genetic algorithm for the evolution of audio. In: Johnson, C., Ciesielski, V., Correia, J., Machado, P. (eds.) EvoMUSART 2016. LNCS, vol. 9596, pp. 63–78. Springer, Cham (2016). https://doi.org/10.1007/978-3-319-31008-4_5
2. Ianigro, S., Bown, O.: Investigating the musical affordances of continuous time recurrent neural networks. In: Proceedings of the Seventh International Conference on Computational Creativity (2016)
3. Ianigro, S., Bown, O.: Exploring continuous time recurrent neural networks through novelty search
4. Eigenfeldt, A., Bown, O., Pasquier, P., Martin, A.: Towards a taxonomy of musical metacreation: reflections on the first musical metacreation weekend. In: Proceedings of the Artificial Intelligence and Interactive Digital Entertainment (AIIDE 2013), Conference, Boston (2013)
5. Van Den Oord, A., et al.: WaveNet: a generative model for raw audio. In: SSW, p. 125 (2016)
6. Kiefer, C.: Musical instrument mapping design with echo state networks (2014)
7. Mozer, M.C.: Neural network music composition by prediction: exploring the benefits of psychoacoustic constraints and multi-scale processing. Connect. Sci. **6**(2–3), 247–280 (1994)
8. Lambert, A.J., Weyde, T., Armstrong, N.: Perceiving and predicting expressive rhythm with recurrent neural networks (2015)
9. Eldridge, A.: Neural oscillator synthesis: generating adaptive signals with a continuous-time neural model (2005)
10. Beer, R.D.: On the dynamics of small continuous-time recurrent neural networks. Adapt. Behav. **3**(4), 469–509 (1995)
11. Bown, O., Lexer, S.: Continuous-time recurrent neural networks for generative and interactive musical performance. In: Rothlauf, F., et al. (eds.) EvoWorkshops 2006. LNCS, vol. 3907, pp. 652–663. Springer, Heidelberg (2006). https://doi.org/10.1007/11732242_62
12. Dahlstedt, P.: Creating and exploring huge parameter spaces: interactive evolution as a tool for sound generation. In: Proceedings of the 2001 International Computer Music Conference, pp. 235–242 (2001)
13. McCormack, J.: Eden: an evolutionary sonic ecosystem. In: Kelemen, J., Sosík, P. (eds.) ECAL 2001. LNCS (LNAI), vol. 2159, pp. 133–142. Springer, Heidelberg (2001). https://doi.org/10.1007/3-540-44811-X_13
14. MacCallum, R.M., Mauch, M., Burt, A., Leroi, A.M.: Evolution of music by public choice. Proc. Natl. Acad. Sci. **109**(30), 12081–12086 (2012)
15. Yee-King, M.J.: An automated music improviser using a genetic algorithm driven synthesis engine. In: Giacobini, M. (ed.) EvoWorkshops 2007. LNCS, vol. 4448, pp. 567–576. Springer, Heidelberg (2007). https://doi.org/10.1007/978-3-540-71805-5_62
16. Bown, O.: Player responses to a live algorithm: conceptualising computational creativity without recourse to human comparisons? In: Proceedings of the Sixth International Conference on Computational Creativity, p. 126, June 2015

17. Bown, O.: Performer interaction and expectation with live algorithms: experiences with Zamyatin. Digit. Creat. **29**(1), 37–50 (2018)
18. Husbands, P., Copley, P., Eldridge, A., Mandelis, J.: An introduction to evolutionary computing for musicians. In: Miranda, E.R., Biles, J.A. (eds.) Evolutionary Computer Music, pp. 1–27. Springer, London (2007). https://doi.org/10.1007/978-1-84628-600-1_1
19. Lehman, J., Stanley, K.O.: Exploiting open-endedness to solve problems through the search for novelty. In: ALIFE, pp. 329–336 (2008)
20. de Castro, L.N., Timmis, J.: An artificial immune network for multimodal function optimization. In: Proceedings of the 2002 Congress on Evolutionary Computation, CEC 2002, vol. 1, pp. 699–704. IEEE (2002)
21. de Franca, F.O., Von Zuben, F.J., de Castro, L.N.: An artificial immune network for multimodal function optimization on dynamic environments. In: Proceedings of the 7th Annual Conference on Genetic and Evolutionary Computation, pp. 289–296. ACM (2005)
22. Abreu, J., Caetano, M., Penha, R.: Computer-aided musical orchestration using an artificial immune system. In: Johnson, C., Ciesielski, V., Correia, J., Machado, P. (eds.) EvoMUSART 2016. LNCS, vol. 9596, pp. 1–16. Springer, Cham (2016). https://doi.org/10.1007/978-3-319-31008-4_1
23. Ableton live, February 2019. https://www.ableton.com/en/live
24. Cycling '74, February 2019. https://cycling74.com

Correction to: Evolutionary Games for Audiovisual Works: Exploring the Demographic Prisoner's Dilemma

Stefano Kalonaris[iD]

Correction to:
Chapter "Evolutionary Games for Audiovisual Works:
Exploring the Demographic Prisoner's Dilemma"
in: A. Ekárt et al. (Eds.): *Computational Intelligence in Music,*
Sound, Art and Design, **LNCS 11453,**
https://doi.org/10.1007/978-3-030-16667-0_7

In an older version of this paper, a footnote was included at the bottom of page 102. This was removed as it no longer linked to the page that it was intended to link to.

The updated version of this chapter can be found at
https://doi.org/10.1007/978-3-030-16667-0_7

Author Index

Printed in the United States
by Baker & Taylor Publisher Services